BOURBON AND RYE

COCKTAILS

BOURBON AND RYE COCKTAILS

→ THE HISTORY OF AMERICAN WHISKEY IN 73 RECIPES ←

ROBERT SIMONSON

PHOTOGRAPHS BY LIZZIE MUNRO

TEN SPEED PRESS
California | New York

CONTENTS

WHAT ARE BOURBON AND RYE, ANYWAY?: A BRIEF HISTORY 15

THE HISTORY OF AMERICAN WHISKEY: A TIMELINE 45

THE RECIPES

BOURBON COCKTAILS

RYE COCKTAILS

INTRODUCTION

It's said that no one knows more about the architecture in Chicago than the city's taxi drivers. Similarly, hop into a cab in Louisville, Kentucky, and you're going to get an earful of whiskey.

My Uber driver's favorite bourbon was Buffalo Trace. I learned this within a minute. I was in town to attend a bourbon convention and to research this book. I had pushed the "quiet ride" button on the Uber app but Mike, the driver, began chatting anyway.

Buffalo Trace came up because I said I was going to be touring the distillery, which is situated in a valley near Frankfort. "No, you're not," said Mike. "Buffalo Trace is under water." I looked out the car window. There was snow on the ground. There had been heavy snowfall the day before, then rain.

"I follow Buffalo Trace," said Mike. He held up a plastic bag. Inside was a bottle. While waiting for a ride earlier, he had darted inside a liquor store and found a bottle of Buffalo Trace for $22. I said that was a good price for that brand; in New York it goes for $37, if you can find it. Mike said Buffalo Trace is almost impossible to find because the Sazerac Company, the distillery's parent company, which is based in New Orleans, ships all its whiskey out of Kentucky. Everybody wants Buffalo Trace.

Mike also liked E.H. Taylor, Jr., and Eagle Rare, both made by Buffalo Trace. I like Eagle Rare myself. I asked if he had seen a bottle of Eagle Rare on the shelves recently. He said no. Bourbon is just too popular these days.

He asked what I liked. I said Michter's; I like their bourbon and rye equally. I said I thought rye, in general, was better if you were mixing cocktails and bourbon was better if you were drinking it by itself. Mike agreed.

Mike said he has a son in Los Angeles who drinks a beverage with THC and mushrooms in it. He shipped some to Mike. "I tried it. It was okay," Mike said. "With bourbon, there's a predictability to it. You know what kind of ride you're going on. With THC, you don't know how long that ride is going to last." It wasn't for him. Mike likes a glass of bourbon.

I told Mike part of the reason I like bourbon is because it comes with a story and a history. THC has no story.

The story of bourbon is a good one. A long one. So is the story of rye whiskey. Both are American stories, as bourbon and rye are our homegrown spirits. This book will attempt to tell those stories.

Currently, those stories have happy endings. Since 2000, bourbon production in Kentucky has increased by 360 percent. Sales reached $5.3 billion in 2024. Every distillery and brand is expanding. Rye whiskey has enjoyed similar and even more surprising growth, going from near extinction at the turn of the century to a time when every distillery producing bourbon must now devote a certain number of days to rye production to satisfy demand. There were once three rye brands on the market; there are now many dozens.

Bourbon and rye owe much of their explosive growth to the cocktail bartending community, which in the aughts began championing the spirits (especially rye) and mixing them into finely tuned drinks. This should be no surprise to anyone. No spirit succeeds in the United States except on the back of a cocktail or two. We are a cocktail country, after all; we invented the cocktail and are justly fond of our creation.

And so, this book on bourbon and rye will tell the stories of those spirits through cocktails. Seventy-three cocktails, to be precise, each drink recipe illustrating a different chapter in the rich, centuries-long history of American whiskey. The chapters are plotted along a timeline, beginning with the building blocks of the spirits (grain, barrels, aging, techniques, geography), then the famous drinks (Old-Fashioned, Manhattan, Sazerac); historical figures from the early days of American whiskey (Elijah Craig, Abe Overholt, Dr. James Crow); mid-history milestones (the Bottled-in-Bond Act, Prohibition, the Taft Decision); storied brands (Buffalo Trace, Maker's Mark, Pappy Van Winkle); and the more recent advances and developments of whiskey production (the Bourbon Trail and craft distilleries). By the end of the book, the tale of America's native whiskeys will have been told.

Note: I sourced these recipes with my wife, Mary Kate Murray, cofounder of my Substack, *The Mix with Robert Simonson*. We used a variety of criteria when choosing which cocktails to include in the book. Some cocktails are classics and were thus natural inclusions. Others made historical sense and tied in neatly to the entry in which they are featured. And still others employed particular bourbons and ryes that are intrinsic to the section in question. A select few are original cocktails that make their debut here and won't be found in any other book. Whenever the creator of the drink is quoted in the headnote, those comments are original to this text.

SWEATER WEATHER

MEAGHAN DORMAN,
RAINES LAW ROOM, NEW YORK CITY (2014)

Meaghan Dorman developed this drink early on during her long reign at Raines Law Room, one of Manhattan's best craft cocktail bars. Her goal was to have the layers of the cocktail take you on a flavor journey. The addition of sherry and orgeat to the base of bourbon and apple cider results in a sophisticated autumnal tipple.

- 1½ ounces apple cider
- 1½ ounces bourbon
- ½ ounce fresh lemon juice
- ½ ounce orgeat
- ½ ounce Palo Cortado Sherry
- Orange twist, for garnish
- Grated cinnamon, for garnish

Combine the liquid ingredients in a cocktail shaker half filled with ice and shake until chilled. Strain into a rocks glass filled with ice. Garnish with the orange twist and a dusting of grated cinnamon.

WHAT ARE BOURBON AND RYE, ANYWAY?

A BRIEF HISTORY

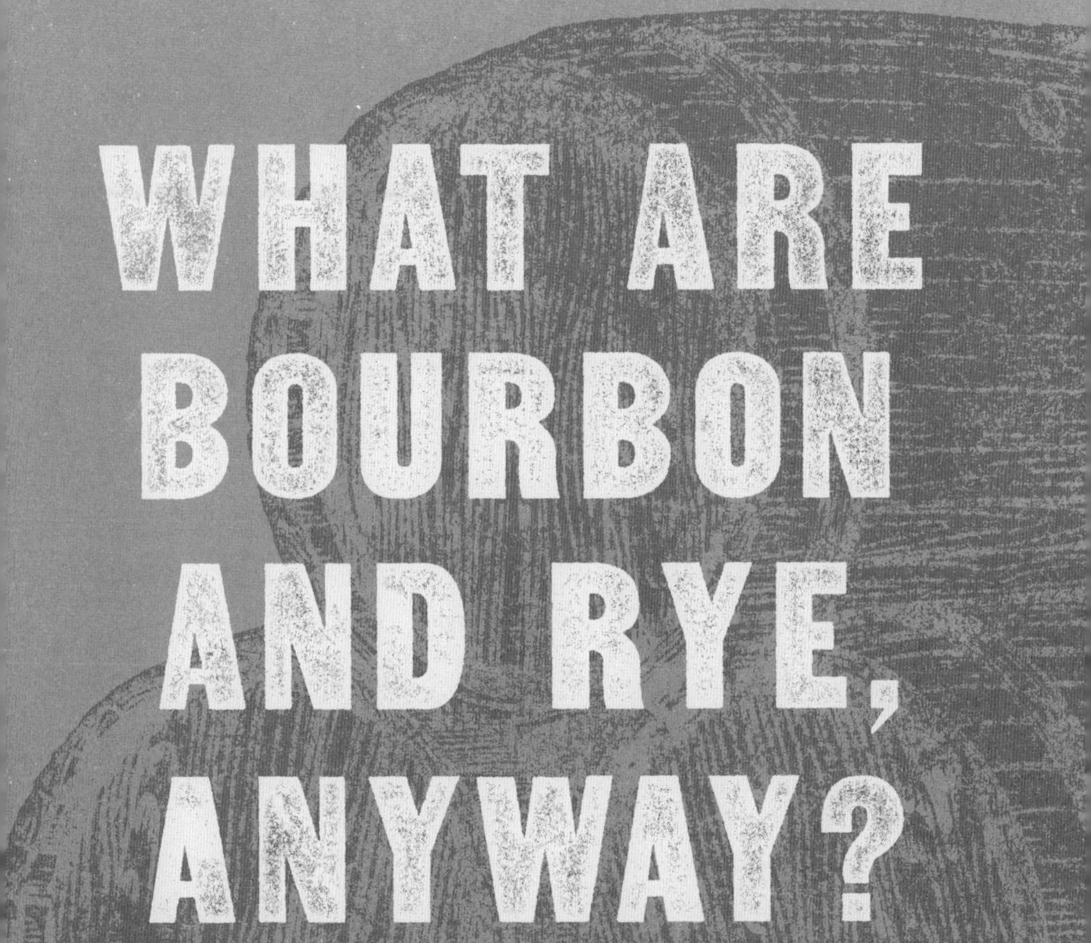

Bourbon and rye are two types of whiskey that, together, can be considered the United States' homegrown spirits. Their histories go back longer than the history of the nation itself. Like all whiskeys, they are made from distilled fermented grain mash. Bourbon is—by tradition and law—predominantly made from corn. Rye is predominantly made from rye grain. Rye leads to a spicier distillate, whereas corn produces a sweeter whiskey. Most bourbon is made in Kentucky, though it can be made anywhere in the United States. Rye used to be made mainly in Maryland and Pennsylvania; today, most of it is produced in Kentucky—though it likewise can be made anywhere in the country, and is. Both spirits, by law, are aged in charred, new oak barrels, vessels that contribute massively to the spirits' colors and flavors. They are then aged in warehouses called *rickhouses,* sometimes for a short amount of time, but typically for many years. When finished and bottled, they don't taste like any of the other renowned whiskeys made around the world.

The fortunes of bourbon and rye have ebbed and flowed dramatically over the past two centuries. The reason you are reading the book in your hands right now is that their fortunes have been booming for the last quarter century and interest in the two spirits has soared. There has never been a better time to be a bourbon or rye drinker and enthusiast, whether you prefer the whiskeys straight or in cocktails.

THE ORIGINS OF THE SPIRITS

The American whiskey revival and the cocktail revival share at least one thing in common in terms of narrative. With both movements, the more we dig into their history, the more we realize we don't know anything. Cocktail historians still don't know with complete certainty why the drinks they study are called *cocktails.* We don't know who mixed the first cocktail, or where. We don't know who created the most famous cocktails, like the Martini, Manhattan, and Old-Fashioned. These are huge questions. And we just don't know.

Similarly, there's a lot about bourbon's origins that we don't know and are unlikely ever to know for sure. For instance, the name *bourbon.* Like *cocktail,* it's a strange name. Scotch whisky is called *Scotch,* because: Scotland. Irish whiskey is called *Irish whiskey.* Rye whiskey gets its name from its source grain. But bourbon? By any logic, it should be called *Kentucky whiskey* or *corn whiskey.* But it's not.

There are theories, of course. The most obvious is that the whiskey was named after Bourbon County. Bourbon County was once a mighty stretch of land. It was established in 1785 and was originally part of Virginia. The name honored the French, in thanks for their assistance during the Revolutionary War. When Kentucky was admitted to the Union in 1792, Bourbon County became part of it. But by then it was shattered into thirty-four smaller counties. The modern Bourbon County is a tiny thing of just under three hundred square miles. Still, some people continue to call the larger area Old Bourbon.

The prevailing idea is that whiskey made in Kentucky was shipped down the Ohio River to the Mississippi out of the town of Limestone (now Maysville) in barrels stamped with its place of origin. Drinkers in New Orleans eventually began calling for "Bourbon" whiskey. But this theory has been often discredited, having little evidence to back it up. However, there are plenty of ads from the 1820s on, in papers from a variety of southern states, advertising the sale or availability of "Old Bourbon County Whiskey" and "Bourbon County Whiskey" from Kentucky. So by then, the terms "Kentucky" and "Bourbon" had become closely linked with whiskey. Some ads used bourbon to distinguish it from other American whiskeys like "Monongahela" rye from Pennsylvania.

There is also another theory that the term "bourbon" was used as a marketing tool by two French immigrants, Jean and Louis Tarascon, shipping merchants who lived in Shippingport, Kentucky, who knew the name would resonate with the French population in New Orleans. Another theory connects the name to Bourbon Street in New Orleans. Either way, New Orleans always seems to be in the mix somehow.

BOURBON COCKTAIL

E. CHISNALL,
***APPROVED COCKTAILS AUTHORIZED BY THE UNITED KINGDOM BARTENDERS' GUILD* (1937)**

Bourbon gets top billing in this sour, which was published in a British bartenders' guild cocktail manual in the 1930s, even though, proportionally, it takes a back seat to the French herbal liqueur Bénédictine.

- 1 ounce Bénédictine
- ½ ounce bourbon
- ½ ounce fresh lemon juice
- Dash of curaçao
- Dash of Angostura bitters

Combine the ingredients in a cocktail shaker half filled with ice and shake until chilled. Strain into a chilled cocktail glass.

BOURBON COUNTY

Kentucky would like you to think that bourbon can only be made in the Bluegrass State, but it doesn't have the law on its side. You can make bourbon anywhere in the United States: Alaska, Montana, Rhode Island, you name it. (Same goes for rye.) But there is an abundance of evidence that would make you think bourbon is exclusively a Kentucky product. In fact, 95 percent of the world's bourbon is produced there.

Kentucky has certain advantages when it comes to bourbon production. There's the climate: The winters are cold and the summers are hot, making for widely fluctuating barrel-aging conditions that result in the particular whiskey flavor profile that the world has come to love. There's the rich soil: Corn and grain grow well there. And under that soil there's the limestone bedrock, through which the water used in making whiskey is filtered. Kentuckians boast about that water and its effects on bourbon the way New York pizza and bagel makers brag about their magic H_2O.

Other states have good limestone water, rich soil, and a similar climate, but none have done a better job of capitalizing on these conditions than Kentucky. In a 1954 ad in the *Lexington Herald-Leader,* J. P. Van Winkle, president of the Stitzel-Weller Distillery, said of whiskey (or his marketing person had him say), "Does it come from the state whose peculiar characteristics of climate, limestone water, grain, closely-guarded family formula and native know-how give it that authentic Kentucky Bourbon flavor?"

And the state has certainly utilized its advantages in terms of number of distilleries (more than one hundred) and distilling capacity (14.3 million barrels aging as of 2024). It's the fruit of a two-century head start. And no one's going to catch up anytime soon.

KENTUCKY BUCK

ERICK CASTRO,
BOURBON & BRANCH, SAN FRANCISCO (2009)

Castro invented this cocktail to shine a spotlight on bourbon, which he felt wasn't getting enough attention in cocktail circles in the early years of the cocktail revival.

2 ounces bourbon

2 ounces ginger beer

¾ ounce fresh lemon juice

½ ounce Simple Syrup (recipe follows)

1 ripe strawberry

2 dashes of Angostura bitters

Strawberry slice, for garnish

Lemon wheel, for garnish

Muddle the lemon juice, simple syrup, and strawberry in a cocktail shaker. Add the bourbon, bitters, and some ice. Shake until chilled. Double strain into a Collins glass filled with ice. Top with the ginger beer. Garnish with the strawberry slice and lemon wheel.

SIMPLE SYRUP

MAKES ABOUT 1 CUP

1 cup granulated sugar

1 cup water

In a medium saucepan, bring the sugar and water to a simmer over medium heat and cook until sugar dissolves. Remove from the heat and allow to cool. Store in an airtight container in the refrigerator for up to 1 month.

CORN

Corn is what bourbon's all about; it's what sets it apart from the world's other whiskeys. By legal definition, bourbon must be made from a grain bill or mash bill (that is, a recipe of different grains) composed of no less than 51 percent corn, and it must be aged in new charred oak barrels. Similarly, rye whiskey must be made of no less that 51 percent rye. By general agreement, corn lends a sweetness to bourbon whiskey that you won't find in the other major whiskey categories.

The corn used for distilling is not the kind you eat on the cob. It is mature corn, left on the stalks until the kernels are dry. Most of what's used in bourbon is a strain called Yellow Dent corn. It got its name from the dent at the end of every kernel. It is not special or rare. Most of the corn grown in the United States is Dent corn. You'll find it in cornmeal, corn chips, and corn syrup. The type is actually a hybrid that goes back to the time when bourbon was just getting going.

An Illinois farmer named Robert Reid developed Yellow Dent in the 1840s. His son, James Reid, entered his corn in the 1893 World's Fair in Chicago and won a blue ribbon. It was popular with farmers because, whereas common corn yields for that time were around twenty-five to thirty-five bushels per acre, Yellow Dent produced more than one hundred bushels per acre. More corn, more money. James Reid described his family's invention in a 1900 ad as "medium in size, medium early, very solid, bright yellow with deep grain, and very popular in the market." If you want to pay homage to the Reids, you can do so. A historical marker was erected in 1955 on the northeast side of Delevan, Illinois. It's about a three-hundred-mile drive from Louisville.

Distillers generally don't discuss corn too much when talking about their product, except to say how much of it goes into the mash bill. Issues like the char level of the barrels, the amount of time the whiskey spends aging in the barrels, and the alcohol level get much more play.

In recent years, young distilleries have begun experimenting with different types of corn; these distilleries include Balcones (Texas), J. Henry and Sons (Wisconsin), New Riff Distilling (Kentucky), Widow Jane (New York), and High Wire (South Carolina). There is a notable difference in taste in these whiskeys. So corn matters. It's just a question of whether you want it in the background or in the forefront.

GURNEY
MASS.
0
5
10
15
20
25
30
U.S
GAUGE CO
N.Y

RYE

Unlike corn, which plays the starring role in the bourbon mash bill, rye has a more versatile and varied role in the American whiskey story. It is often seen taking a supporting role in bourbon recipes. But it is also a star player in its own right, being the marquee grain in the rye whiskeys made in Kentucky (where most rye is made these days) as well as in Pennsylvania and Maryland ryes, bygone styles that were dominant until Prohibition and are only now making a comeback.

Corn is Tom Cruise, always getting top billing. Rye is Philip Seymour Hoffman, shining equally in starring and supporting parts.

Just as it makes sense that corn whiskey flourished in Kentucky in the nineteenth century because the crop grew well there, it makes sense that rye whiskey flourished in Maryland and Pennsylvania. Rye grows well in poorer soils and can withstand cold weather better.

The two states historically produced distinct styles of rye whiskey. Pennsylvania rye—also called Monongahela rye after Western Pennsylvania's Monongahela Valley—had no corn whatsoever in its mash bill. It was composed of only rye—usually about 80 percent—and malted barley. Maryland-style rye did have some corn in it, anywhere from 20 to 30 percent.

When the Maryland and Pennsylvania rye industries died out after Prohibition, Kentucky took over production of most of the rye whiskey in the country, though that wasn't all that much. They did it the Kentucky way, however, using a good deal of corn in their recipes and just enough rye to qualify as rye whiskey under Federal law. Their ryes tasted a lot like high-rye bourbons. Imagine ordering Buffalo wings at a bar. You order Kentucky rye if you want "mild"; Maryland rye if you want "medium"; and Pennsylvania rye if you want "spicy."

Before the rye revival took hold in the early twenty-first century, the Kentucky distilleries that did make rye distilled it only one day a year; the demand was that low. Production schedules have since increased, which is probably not much welcomed by distillery workers. Distillers often talk about the difficulty in distilling rye, as opposed to corn. As a grain, it's thick and gummy and makes distilling equipment difficult to clean.

R & D (RYE AND DAMSON)

DAVE ARNOLD,
BAR CONTRA, NEW YORK CITY (2024)

This is one of the simplest drinks ever devised by New York bartending wiz Dave Arnold, who is known for his use of complex, cocktail-building techniques. "I decided to call it R&D as a double entendre—Rye and Damson, and a joke about research and development, because this drink had none," said Arnold. "It just happened. It is the only drink I have ever made that requires no prep or equipment of any kind."

1½ ounces rye, preferably Rittenhouse

1½ ounces Averell Damson Gin Liqueur

Combine the ingredients in a mixing glass half filled with ice and stir until chilled. Strain into a chilled Nick & Nora glass or smallish coupe.

WISCONSIN COCKTAIL

HYMAN GALE AND GERALD F. MARCO,
THE HOW AND WHEN (1937)

I love this drink. The 1937 instructions in *The How and When* say to shake this drink, but that's nonsense. Stir it. The green maraschino cherry, however, is a must. It is that garnish alone that makes me believe this might have once been a popular cocktail in Wisconsin. Wisconsin is a goofy state where garnishes are concerned. Reese and Roland make versions of the green maraschino cherry you can easily buy online.

1¾ ounces rye

¼ ounce Marie Brizard Apry liqueur

Green maraschino cherry, for garnish

Combine the liquid ingredients in a mixing glass half filled with ice and stir until chilled. Strain into a chilled cocktail glass. Garnish with the green cherry.

WHEAT

Wheat is the third grain in the bourbon story, not used as a supporting grain as frequently as rye, but often stealing the show through its cameos in some of the world's most renowned bourbons. Maker's Mark, one of the most famous bourbons in history, is what's known as a "wheated bourbon." Pappy Van Winkle, the most sought-after bourbon in the world, is also a wheated bourbon. Other famous wheated bourbons include Old Fitzgerald, Rebel Yell, and W. L. Weller.

These are not to be confused with wheat whiskey, which must contain 51 percent wheat in its mash bill and is not bourbon at all.

Wheat is a game changer in terms of flavor. Once you try a wheated bourbon, you will never mistake it for a bourbon that has rye in its recipe. Wheat is a softening and smoothing agent. It doesn't have the bite high-rye bourbons do, and that is an appealing aspect to many drinkers.

Wheated bourbons used to be relatively rare. This is mainly because wheat was usually reserved for baking, the grain being prized higher than rye for that purpose. But in recent years they have become more common, even trendy. Wheated bourbon brands launched in recent years include Green River, Penelope, Larceny, Wilderness Trail, and one of Heaven Hill's Grain to Glass series.

THE CARPETBAGGER

CHARLES HARDWICK,
THE BLUE OWL, NEW YORK CITY (2008)

Sorel, a bright red liqueur made with hibiscus, ginger, cassia, nutmeg, and cloves, lends this relaxed bourbon sipper a complex spicy undertone.

2 ounces wheated bourbon, preferably Maker's Mark

½ ounce orange curaçao

½ ounce Sorel Hibiscus Liqueur

2 dashes of orange bitters

Orange twist, for garnish

Combine the liquid ingredients in a mixing glass and stir until chilled. Strain into a rocks glass over one large ice cube. Garnish with the orange twist.

THE SOUR MASH METHOD

In 2008, Woodford Reserve released a limited-edition bottling, as part of its Master's Collection series, called Woodford Reserve 1838 Sweet Mash. Which left a lot of bourbon novices scratching their heads and asking, "What the heck is sweet mash?"

To understand the audacity of Woodford Reserve's experiment you had to know that almost every bourbon on Earth is made using the sour mash method, something that's as important to making good bourbon as the mix of grains, the use of oak barrels (see page 38), and the charring of those barrels.

Remember the early days of the Covid-19 pandemic, when everyone, stuck at home with lots of time on their hands, started making sourdough bread? To make sourdough, you need a living culture of yeast and bacteria called a sourdough starter, a sort of spark that ignites the bread-making fire. Sour mash is sort of like a bourbon starter. Distillers save the detritus—grain, dead yeast, water—from a previous distillation and fold it into a new batch of bourbon. The dead yeast is a good source of food for the live yeast; the sour mash helps control the acidity of the whole mash; and sour mash helps provide a consistent character from batch to batch. And, unlike wine, consistency is the whole ballgame in bourbon.

It wasn't always that way. Long before secondhand mash became the industry standard, the liquor was made using fresh ingredients only—a sweet mash. Woodford's fleeting product revived this erstwhile method. "The process is riskier, which makes it more expensive," Woodford master distiller Chris Morris said at the time.

Sour mash isn't a special style of American whiskey, as some think. That misconception is primarily the fault of Jack Daniel's, which put "sour mash whiskey" on its label—a phrase that makes about as much sense as if they distinguished themselves by saying "barrel-aged whiskey."

Most people give credit to Dr. James C. Crow (see page 57) for standardizing the sour-mash process in Kentucky in the early 1800s, which is not to say he invented the method. Early ads for his product refer to the "James Crow formula."

BARREL HOPPER

JEFFREY KNOTT,
TARTAN HOUSE, LOUISVILLE, KENTUCKY (2024)

"I wanted a refreshing, but boozy, drink in the vein of the Frozen Irish Coffee from Erin Rose in New Orleans," said Knott. "And while the Barrel Hopper is not frozen, it's almost as refreshing in the New Orleans heat."

2 ounces Elijah Craig Small Batch Bourbon

1 ounce Fernet Branca Menta

1 ounce Reàl Coconut Cream

6 dashes of Scrappy's chocolate bitters

Grated cinnamon, for garnish

Combine the liquid ingredients in a cocktail shaker half filled with ice and shake until chilled. Strain into a rocks glass over crushed ice. Grate the cinnamon on top.

STILLS

Stills are the ornate contraptions that turn what are essentially the fermented grains used in beer into whiskey. There are two main types: the pot still and the column still. The pot still was the more common until an Irishman named Aeneas Coffey patented the column still in 1830. The names say it all: The pot still looks like a big, squat pot with a long, pointy lid and the column still looks like a long, tall column. The differences go on. Pot stills are usually made of copper; column stills can be copper or stainless steel or both. The pot still is more work; you have to clean it after every distillation. The column still is also called a continuous still, because it can distill continuously. The column still is thus much more efficient, and became the norm in the years after the Civil War. Debate continues today as to which is the better method. Purists tend to argue that pot stills produce a more flavorful, complex, richer distillate and that column stills strip away many of those qualities. And yet, almost all of the great bourbons and ryes we're drinking come off column stills, so they can't be all that bad.

Most of the stills used by bourbon and rye distilleries in the United States are made by Vendome, a family-owned company based in Louisville that has done business for more than one hundred years. It wasn't founded by a family named Vendome, as you might expect, but by one Elmore Sherman. But if you go to some smaller craft distillers, you'll see stills made by Carl, a German manufacturer. Additionally, many modern stills are hybrids of pot and column stills.

BARRELS

Ironically for a spirit that is emblematic of the South—a region of the United States that has always embraced a persona of independence and rebellion—bourbon is kept on one of the tightest and shortest leashes ever devised by the federal government. It is, in fact, the most regulated spirit in the world. There are laws upon laws upon laws dictating how you can or can't make it.

One of those laws is about barrels. Scotch and Japanese whiskeys, as well as rum and various brandies, are aged in all kinds of barrels—old, new; big, small; made of various kinds of wood. Bourbon must be aged in newly charred oak barrels. For would-be bourbon makers, the choices are: take it or leave it.

Oak is the final ingredient in any bourbon or rye. It does its work post-distillation, when the spirit is poured into a barrel, but its effect is outsized. Without barrels, we would all be drinking moonshine. Charred white oak barrels lend bourbon and rye all of their color and much of their flavor. If you think your bourbon or rye is "smooth," you can thank the barrel. That's what barrels do for whiskey.

The law only requires oak barrels, but most distilleries opt for white oak, which grows in the northern United States and Canada. White oak is favored because of the natural chemical compounds it contains—elements that lend bourbon

NYDC

its distinctive flavors of vanilla, cinnamon, caramel, and coconut. The required charring of the inside of the barrels creates a charcoal layer that naturally filters out impurities in the whiskey. The charring also caramelizes the wood, imparting attractive flavors. During warm weather, the whiskey expands and pushes through the layer of char to the wood beneath, gathering up all the tasty aspects of that wood. When the weather gets cold, the whiskey contracts, taking all that acquired flavor with it.

And that process is repeated again and again for as long as the whiskey stays in the barrel. In and out, in and out. It's the simplest of machines, pushed along by a brief marriage with fire and the mood swings of Mother Nature.

If you think the choice of barrel isn't all that important, try an experimental "bourbon" that has been aged in sherry butts or French Limousin oak, the kind used for Cognac. It just doesn't taste the same. In fact, it hardly tastes like bourbon at all. Having the proper barrels is so important to the industry that some distillers once owned their own cooperages.

The American whiskey industry's standard barrel size settled at 53 gallons in the years after World War II. Why? Because that was the maximum size that would fit in the average aging warehouse at the time. A filled barrel weighs roughly five hundred pounds.

The whiskey world was reminded of the importance of barrel size when the craft distillery boom dawned in the early aughts. There's no law that says what size your oak barrel has to be. So many of these new distilleries opted for smaller vessels, often *much* smaller. These tight containers exposed the whiskey to more wood surface area and sped along the aging process, giving the newbies something to sell quicker. It was a cheat, basically. Color came quickly to the liquid. Mature flavor was considerably harder to come by.

Small-barrel whiskey took plenty of knocks for its sped-up methods. One whiskey writer published an entire booklet called *Small Barrels Produce Lousy Whiskey*. Point taken.

Many bourbon and rye brands tout the advanced age of their special releases. But unlike Scotch, which ages more slowly owing to the cooler climate in Scotland, double-digit age statements don't necessary result in better whiskey in Kentucky or elsewhere in the United States. Most bourbon and rye distillers consider six to ten years to be the sweet spot for American whiskey. After that, the rough wood notes imparted by the barrels tend to take over the flavor. So, when faced with a 15-, 20-, or 25-year-old bourbon or rye selling for a high price, buyer beware. Your money is likely better spent on a younger bottle.

CLYDE COMMON BARREL-AGED COFFEE MANHATTAN

JEFFREY MORGENTHALER,
CLYDE COMMON, PORTLAND, OREGON (2016)

In the late aughts, at his bar in Portland, Morgenthaler became the king of the barrel-aged cocktail, kicking off a trend that would quickly gird the globe. This is one of his recipes from that time. There is a time investment, but it is worth it, and the effort yields many evenings of refreshment. Small barrels can be purchased online and, occasionally, sourced from local microdistilleries.

12 ounces whole medium-roast coffee beans

½ gallon rye (roughly 2½ [750 ml] bottles)

¼ gallon sweet vermouth (roughly 1¼ [750 ml] bottles)

¾ ounce Angostura bitters

Brandied cherry, for garnish

In a 3-liter oak barrel, add the coffee beans. Seal and set aside to rest for at least 1 month. At the end of the month, empty the beans from the barrel (this takes some time and patience, and having another person help turn the barrel upside down to achieve this is beneficial).

In a separate container, stir together the rye, sweet vermouth, and bitters. Add the mixture to the barrel, seal it, and age, at room temperature, for at least 2 months, checking the flavor frequently. When ready (it should taste like a smooth Manhattan), empty the barrel and fine-strain the Manhattan to remove any barrel char or sediment. Pour into individual (750 mL) bottles, seal, and label. (This recipe can be scaled down for smaller barrels.)

To serve, add 3 ounces Coffee Manhattan to a mixing glass half filled with ice and stir until cold. Strain into a chilled cocktail glass and garnish with the brandied cherry.

RICKHOUSES

The bourbon business has a special term for warehouse. It's *rickhouse.*

Rickhouses are where the barreled bourbon is aged. They are therefore a very influential force on the sort of liquid that eventually pours out of the bottle. Rickhouses take many forms. They were once made of wood—the racks, or "ricks," that give the structure its name—and have many windows, which allows for the easy flow of air in and out of the building. Some from the mid-twentieth century were encased in brick. More modern ones are made of cement and metal.

Rickhouses are typically many stories high, with levels of barrels reaching up to the ceiling. But not always. Four Roses, for instance, uses a rickhouse that is only one story high. Sometimes theses barrels are rotated to even out the effects of the varying climates found on different floors. Barrels on the top level and near the walls are exposed to more heat than those at the bottom and middle of the rickhouse, and therefore age differently.

Other distilleries choose to leave the barrels where they are for the entirety of the aging process and to work out their differences later during blending. Some distilleries chose not to leave the aging process entirely up to Mother Nature, instead controlling the temperature in their rickhouses through heating and cooling systems. Where the rickhouse is built—valley, hill, sunny place, windy site—also has an impact on the whiskey it produces. The biggest distilleries own multiple rickhouses, often numbering in the dozens. Frequently, they stand in different towns.

There's a company in Kentucky called Buzick Construction that specializes in building rickhouses and has built many of the ones you see dotting the roadsides in that state. Along with the still-makers Vendome Copper and Brass Works and the coopers Independent Stave Company, it is one of the symbiotic companies that operate hand in glove with the American whiskey industry.

With the bourbon and rye boom of the last quarter century, rickhouses have sprouted like mushrooms across Kentucky and elsewhere. They can be very romantic places for the bourbon tourist. Few sights tingle the senses more than the first walk down the aisle of a cool, dark rickhouse, with aromatic barrels climbing to the ceiling on both sides and smelling of whiskey and wood.

There is less romance when the buildings are wholly cement or metal. Microdistilleries possess almost no poetry at all, as they can't afford a separate warehouse and simply store their barrels in a corner of the same building where they make the whiskey.

As the bourbon world heated up in the aughts, one whiskey-focused cocktail bar in San Francisco capitalized on the renewed interest, calling their bar Rickhouse.

TIME AND A HALF

CHARLOTTE DANIEL,
THE KLOCKE ESTATE, HUDSON, NEW YORK (2024)

"I'm a bourbon girl," said Charlotte Daniel, who also co-owns Hemlock, a cocktail bar in Catskill, New York. "I come by it honestly from my father who's an Atlanta native. He's been drinking exactly one bourbon on the rocks with a splash of water my whole life. Never rye. Never Scotch. Only bourbon. As I got older it was something we bonded over. The Time and a Half is a punchy yet smooth mix of a bit of crème de cacao for some roundness, Demerara to kick off some of the heat, Old Forester 86 bourbon (of course), and Roger Groult Calvados as a nod to the Klocke Estate's brandy offerings."

1½ ounces Old Forester 86 proof Bourbon

¾ ounce Calvados, preferably Roger Groult

¼ ounce Demerara Syrup (recipe follows)

¼ ounce crème de cacao

Absinthe

Combine all ingredients except the absinthe in a mixing glass half filled with ice. Stir until chilled. Strain into a cocktail glass. Using an atomizer, generously spritz absinthe over the surface of the drink.

DEMERARA SYRUP MAKES ABOUT 1 CUP

1 cup Demerara sugar

1 cup water

In a medium saucepan, bring the sugar and water to a simmer over medium heat and cook until sugar dissolves. Remove from the heat and allow to cool. Store in an airtight container in the refrigerator for up to 1 month.

THE HISTORY OF AMERICAN WHISKEY

A TIMELINE

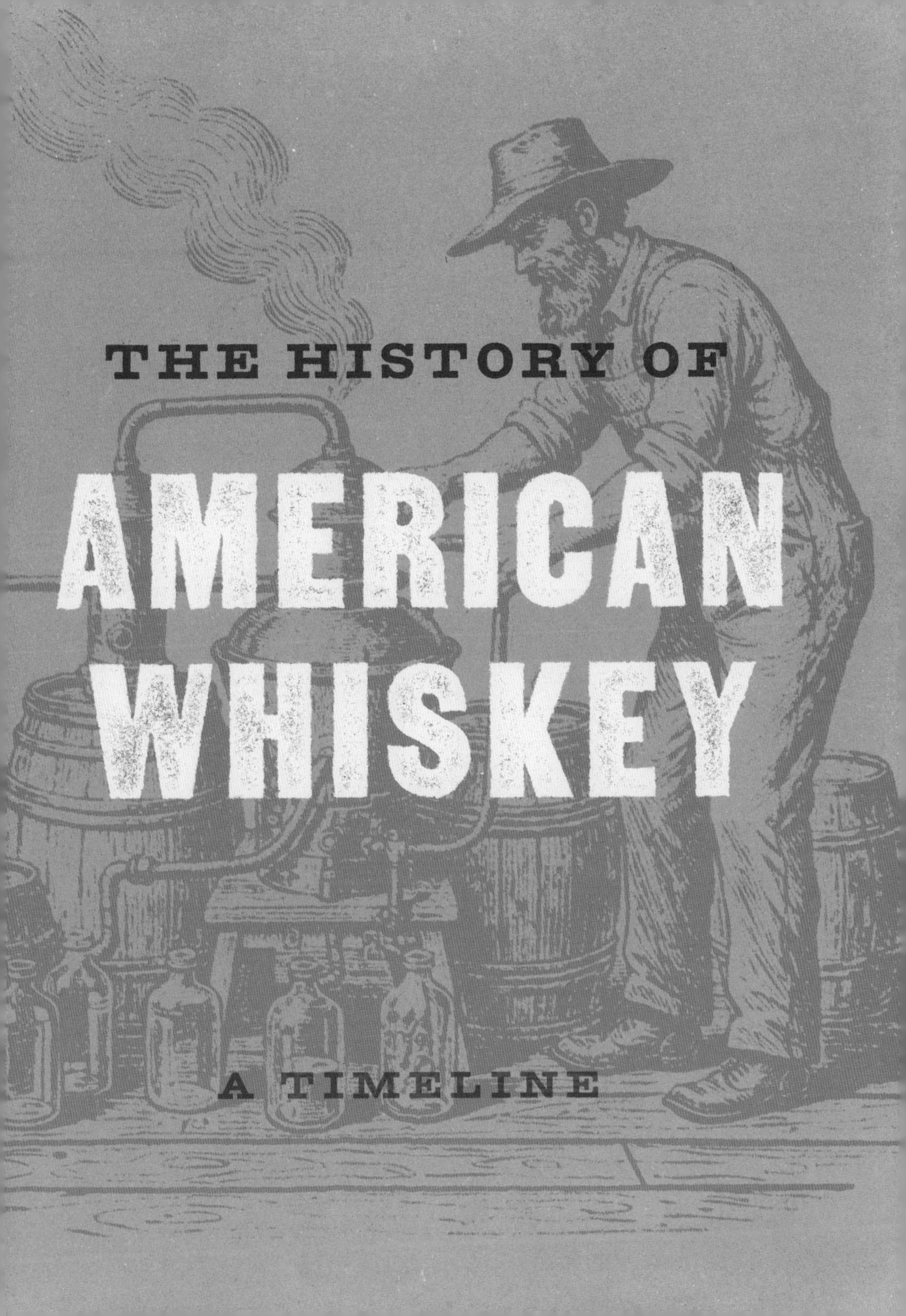

1785

BOURBON COUNTY IS ESTABLISHED AS PART OF VIRGINIA.

1789

ELIJAH CRAIG

Bourbon is a success story. As such, it has many fathers. One of the names that comes up regularly as a progenitor of the spirit is Elijah Craig, who, drolly enough, was a man of the cloth, a Baptist minister. He was born around 1738 in Virginia, ordained in 1771, and, around 1782, followed his brother Lewis, who was also a preacher (as was a third brother), to what is now Kentucky. There he lived until his death in 1808, when the nation was just over thirty years old and the bourbon whiskey tradition was in its infancy.

Craig did a lot. He was an educator, establishing a school in Fayette County, founded several mills, preached at several churches, and served as fire chief. He was an accomplished man, one of the richest in his county, and owned four thousand acres at his death. It is ironic that he is best remembered today for the thing he probably didn't do, which is invent bourbon. He did distill. There's no doubt about that. Craig founded a distillery in Georgetown in 1789. He is regularly credited with being the first to age whiskey in charred oak barrels—newspapers during the latter half of the twentieth century repeatedly stated this as fact—but there is little evidence to back up this claim, and most bourbon historians doubt it.

Nonetheless, Craig's name regularly lends a smile to bourbon lovers' lips because it is attached to one of the better affordable whiskeys out there. Along with Evan Williams (see page 126), the Elijah Craig brand is the foundation of the Heaven Hill distillery (see page 104). As Chris Morris, master distiller of Brown-Forman (see page 134), told me, the two labels "combined over a generation to make Heaven Hill a bourbon powerhouse." It was released in 1986, at a ridiculously low price, and at the then-unheard-of age of twelve years. That expression is no longer made, but it is fondly remembered. Today's Elijah Craig is between eight and twelve years old. There are also older expressions of Elijah Craig 18-year-old and 23-year-old on the market that fetch a pretty penny.

PARTY HAT

LINDSAY MATTESON,
AMOR Y AMARGO, NEW YORK CITY (2015)

Matteson told me that this drink was designed as a more accessible Boulevardier, with a lot of orange flavor. "Orange and bourbon is such a delicious combination, and orange makes this cocktail about as approachable as you can get for a stirred whiskey cocktail. The Cocchi Americano adds that mouth-watering quality and a lot of gentle citrus. Montenegro is very approachable and has a lot of orange flower and cucumber notes that fit perfectly, with the cucumber amplifying the juicy flavor and the orange taste and aromatics really making the orange flavor pop. Finally, the orange bitters add some balance and structure, while echoing the orange notes."

1½ ounces Elijah Craig bourbon

¾ ounce Cocchi Americano

¾ ounce Amaro Montenegro

2 dashes of orange bitters

Orange twist, for garnish

Combine the liquid ingredients in a mixing glass half filled with ice and stir until chilled. Strain into a double Old-Fashioned glass filled with ice. Garnish with the orange twist.

1791–1794

THE WHISKEY REBELLION

The first important chapter in the story of American whiskey is very on-brand. It included a rebellion, anti-government fury, guns, the military, and the father of our country, George Washington. What's more American than that?

The United States' first secretary of the treasury, the wunderkind, big-government advocate, and future subject of a Broadway musical, Alexander Hamilton, had to pay off debts from the Revolutionary War. Taxes were needed. So, in 1791, with the country only fifteen years old, he decided to tax whiskey, a product many frontier farmers made from their surplus grain, both to enjoy themselves and to sell to others. It was the first tax imposed on a domestic product in the United States and the first instance in the long American tradition of using "sin taxes" to pay for a lot of other things.

No American liked taxes—that's what the revolution was about after all, anger over taxes—so the farmers bucked. They whipped, tarred, and feathered tax collectors from the East. Tensions escalated for years until, in 1794, Hamilton asked his boss—who was still best known as a general—to personally take the army to western Pennsylvania and remind those rebellious hicks who was in charge. Washington did, becoming the first, and so far only, sitting commander in chief to lead an army into battle. It worked. The uprising was quelled.

The whiskey that Pennsylvania farmer-distillers were making was mainly rye. There's a popular myth that the unhappy farmers picked up and moved from the commonwealth of Pennsylvania to the frontier land of Kentucky, which was not yet a state and where, presumedly, they could do what they wanted and began to make corn-based bourbon instead of rye. This isn't true. The rebel leaders largely escaped to Louisiana. Whiskey production continued in western Pennsylvania. Besides, the tax applied to Kentucky as well; it just wasn't as strictly enforced there.

The Whiskey Rebellion is still remembered in the area, if only whimsically. During the craft distillery boom of the 2010s, a Pittsburgh distillery named itself after Phillip Wigle, a rebel who was arrested, tried, sentenced to hang, then pardoned by Washington. Washington, ironically, distilled rye whiskey at Mount Vernon.

The whiskey tax was repealed in 1802 by the agrarian-minded Thomas Jefferson, who was probably thinking less of the poor farmers than of a way to stick it to his lifelong rival, Hamilton. Today, there is a bourbon and a rye named after Jefferson. But there's also a New York rye named after Fort Hamilton, which was named in honor of Hamilton. There's more irony for you.

REMEMBER THE MAINE

CHARLES H. BAKER, JR.,
THE GENTLEMAN'S COMPANION (1939)

This cocktail and the works of Baker, a dynamic mid-twentieth-century globetrotting journalist and sybaritic gatherer of drink recipes, would have probably been lost to time if not for the efforts of Brooklyn bartender and historian St. John Frizell. Thanks to Frizell, the Remember the Maine is now a recognized classic.

2 ounces rye

¾ ounce sweet vermouth

Barspoon of Cherry Heering

Dash of absinthe

Lemon twist, for garnish

Combine the liquid ingredients in a mixing glass half filled with ice and stir until chilled. Strain into a chilled coupe. Garnish with the lemon twist.

1792

KENTUCKY JOINS THE UNION.

1806

THE WHISKEY COCKTAIL

The Whiskey Cocktail is one of the first American mixed drinks to use American whiskey; one of the simplest; and still one of the most popular. You know it today by the name Old-Fashioned (see page 80). But for its first several decades, Whiskey Cocktail was enough of a descriptor.

The formula adheres to the first published definition of a cocktail from 1806 as a mixture of whiskey, sugar, bitters, and water, eventually served over ice, but in its first flush of popularity, just served as is. Early versions were probably made with rye whiskey, with bourbon stepping in in the late nineteenth century. Celebrity bartender Jerry Thomas put the Whiskey Cocktail in his famous 1862 cocktail book; he didn't specify which whiskey to use. It was the "Improved" version of this cocktail that led to the coining of the Old-Fashioned name, which supplanted it.

WHISKEY COCKTAIL

JERRY THOMAS,
THE BAR-TENDER'S GUIDE: HOW TO MIX DRINKS, A BON-VIVANT'S COMPANION **(1862)**

The instructions here, complete with cracked ice and shaking, are per Thomas, and are reprinted in the interest of historical drinking. But if you want to stir this cocktail, you're within your rights.

2 ounces rye or bourbon

Barspoon of gum syrup or Simple Syrup (see page 23)

2 dashes of Angostura bitters

Lemon peel, for garnish

Combine the liquid ingredients in a cocktail shaker filled with cracked ice and shake until chilled. Strain into a chilled rocks glass. Garnish with the lemon peel.

1810

OLD OVERHOLT RYE

An hour's drive south of Pittsburgh, there is a small village named West Overton. It was established in 1803. If there is an Overton village proper, I can't find it on the map. I also can't find a population count. That's probably because no one lives there. But there are buildings. Quite a few, actually. One looks like a mansion, another like a barn, another like a distillery, others like workers' cottages.

They *were* a mansion, a barn, a distillery, and workers' cottages. West Overton was, in fact, a company town. Today, it may very well be the best-preserved nineteenth-century company town in the United States. The product the company made was initially whiskey, until coke and coal production took over later on. You get a hint as to which whiskey that was when you learn West Overton was founded by a group of Mennonites led by one Henrich Oberholtzer; adopted name: Henry Overholt.

West Overton Historical Village is the greatest whiskey heritage spot that isn't on the Bourbon Trail. It's the birthplace of Old Overholt, one of the most famous liquor brands in American history and the oldest continually produced rye whiskey around. The collection of nineteen historic buildings across forty acres now operates as a living museum. There are structures that were once the general store, the cooperage, and the post office. The barn is the largest standing brick barn left in Pennsylvania. Taken all together, it is the only complete pre–Civil War village still intact in Pennsylvania.

Jim Beam (see page 98) is the current owner of Old Overholt. The brand has been produced and bottled by the bourbon giant for many years in Clermont, Kentucky. Beam acquired the brand—which was founded way back in 1810 by Henry's son Abe Overholt—in 1987. The new museum is the result of a collaboration between Beam's mighty parent company, Suntory, and the West Overton Village & Museum, a nonprofit.

The Overholts lived pretty well in a handsome old Federal-Style house built in 1838, a three-story, four-chimney brick building with a long wooden porch. It sits atop a slight incline with a clear view of the distillery and all the out buildings. There is a pantry where Abe was in the habit of enjoying a daily dram of his spirit.

In 1922, Abe's great-granddaughter, Helen Clay Frick, acquired the homestead, opening it as a museum in 1928. Yes, Frick. To liquor-history mavens like myself, the West Overton property is Overholt Land. It's all about the rye. But to most historians, it is known as the birthplace of industrialist Henry Clay Frick. To the right of the

Overholt home is a small stone building that was once Abe's spring house. On it is a modest plaque that reads, "Henry Clay Frick. Born here December 19, 1849."

There is no clearer illustration of the angel-devil nature of American capitalism than the direct line between Abraham Overholt and Henry Clay Frick. One was a devout Mennonite who lived a modest life and tried to practice honesty in his business dealings. The other was one of the fiercest, most evil-minded robber barons who ever breathed.

But if it weren't for Frick and his lucre, it's possible West Overton Village would not exist today. Frick left his fortune to his daughter, Helen, who purchased the property as a way to memorialize her father. Following a recent renovation, it reopened to tours in 2013.

Frick sold a portion of Overholt to his plutocrat pal Andrew W. Mellon. After Frick died, Mellon had to sell it because he was then Secretary of the Treasury. Also, it was Prohibition. After that, the brand shuttled around between various owners, eventually becoming the last big rye brand standing, steadily declining in quality and popularity along the way, until Beam acquired it in 1987. By that time, it was sold at a lowly 40 percent alcohol; it used to be a bonded product (see page 84). Still, during the early years of the cocktail renaissance, Old Overholt was a bartender favorite, mainly because of its heritage and because it was one of the few ryes out there.

Over the past few years, a contingent at Beam has been trying to resurrect Overholt's reputation by various measures. They released a bonded version; in 2020, they brought the ABV of the standard expression up to 86 proof and restored the age statement of 4 years; and there were limited releases of 10- and 11-year-old Cask Strength Rye and a version bottled at 114 proof. Each of these moves helped to restore integrity and full-bodied flavor to the brand and rid it of its bottom-shelf reputation.

Overholt's newest release is A. Overholt, which uses the same mash bill Abraham Overholt used in the nineteenth century: 80 percent rye grain and 20 percent malted barley. (The current Old Overholt Rye uses a good portion of corn.) At this pace, it won't be long before they make Overholt rye in West Overton again.

MOTT & MULBERRY

**LEO ROBITSCHEK,
NOMAD BAR, NEW YORK CITY (2013)**

Leo Robitschek, then the bar manager at the NoMad bar in Manhattan, created this cocktail when I, as the resident cocktail writer at *The New York Times*, asked him to come up with a new drink for the holidays. "The best cocktails always tell a story and evoke sense memory," Robitschek told me. "There are no cocktails that are more successful at this than holiday cocktails." With his Mott & Mulberry, named for two streets in Little Italy in homage to its Italian and American ingredients, Robitschek aimed to create a drink that was festive, like a hot spiced cider, but avoided the rich excesses of some traditional holiday tipples. It is basically a whiskey sour armed with two secret weapons: the distinctly sweet-tart flavor of Honeycrisp apple juice and Amaro Abano, which the bartender called a "spice bomb, adding cinnamon and clove notes with a pleasant bitterness."

- 1 ounce Old Overholt Rye
- 1 ounce Luxardo Amaro Abano
- ¾ ounce Honeycrisp apple cider or tart apple juice
- ½ ounce fresh lemon juice
- ½ ounce Demerara Syrup (see page 43)
- Thin apple slices, for garnish

Combine the liquid ingredients in a cocktail shaker half filled with ice. Shake until chilled. Strain into a rocks glass over ice. Garnish with the apple slices, skewered at one end with a toothpick and spread like a fan.

1812

OSCAR PEPPER AND DR. JAMES CROW

A candidate for the most important patch of land in bourbon country is the site of the former Pepper distillery near Versailles, Kentucky, where the Woodford Reserve distillery (see page 144) now operates. Here, a distillery was founded by Elijah Pepper in 1812. He passed it on to his son Oscar, who hired one Dr. James Crow, a Scot from Inverness who was educated in medicine and chemistry and who moved to Woodford County in the 1820s.

Crow introduced practices that became standard operating procedure, including the sour mash process (see page 34) and routine barrel aging. The former method, in which a bit of the previous batch of mash is transferred to the next, ensured consistency of flavor and quality. No one lives today who tasted the good doctor's joyous juice. Still, bourbon probably wouldn't be what it is if it weren't for Dr. James Crow.

Crow died in 1856 at age sixty-seven, but Oscar Pepper continued to bottle the popular whiskey called Old Crow. Pepper died in 1859, after which his distillery and stock were acquired by a firm called Gaines, Berry & Company, later known as W. A. Gaines and Co. They kept making Old Crow the Crow way, with the help of his former assistant William F. Mitchell as distiller. They also registered the trademark Old Crow Whiskey in 1870. In 1868, the company bought land three miles down the road and built the Old Crow Distillery. The whiskey was made there, with a break for Prohibition, until 1987.

Dr. Crow's and Old Crow's reputations grew over the decades. The *State Journal* of Kentucky said in 1960, "Every so often the tide of history brings together the right man and the right moment and an event takes place that stands out for years to come. It was indeed a fortunate moment for Kentucky and the entire liquor industry when James Crow decided to leave Scotland, and eventually to settle in what was then a primitive backwoods section of Kentucky."

Old Crow was a popular bourbon for a long time, stretching into the mid-twentieth century, when National Distillers owned the brand. But quality began to decline. In the 1950s, the proof was lowered to 86, though the traditional bonded version remained available. Jim Beam has made Old Crow since 1987, which is to say it churns out a bottom shelf item that it has decided to call Old Crow. Ironically, the bourbon that was once more expensive than any other in the good doctor's day is now cheaper than almost any other.

1862
WHISKEY SOUR

In popularizing bourbon and rye as mixing spirits, the Whiskey Sour is as important as the Old-Fashioned and the Manhattan, if far less glamorous. It was an innovation, however. Once upon a time, mixing spirits and citrus was not a thing one did with cocktails. Then a New Orleans bartender named Joseph Santini whipped up a drink called the Brandy Crusta, a Cognac cocktail supported by lemon juice, curaçao, and maraschino liqueur, garnished with a long lemon twist that snakes inside the glass and with a sugared rim.

It was this recipe, not one for a Whiskey Sour exactly, that bartender Jerry Thomas put in his 1862 cocktail book. (Other spirits like brandy and gin got their own sour, using that term.) After that, the world was flush with the genre of drinks now called *sours,* some of which took on the actual name of *Sour*—like our friend here—and some of which didn't, like the Daiquiri and the Sidecar.

The Whiskey Sour didn't have a roller-coaster career like some of its whiskey cocktail brothers, changing form every generation or two. It pretty much stayed a straight-ahead whiskey, lemon juice, and sugar affair. Its main point of contention, interpretation-wise, was the sometimes-appearing fourth ingredient: an egg white. It was the Neil Young to the Sour's Crosby, Stills & Nash—on some albums, not on others; sometimes touring with the group, sometimes sitting it out.

Egg whites, as they apply to sours, follow an inconsistent logic. In some classic sours, like the Pisco Sour and Clover Club, they are used as a matter of course. In others, like the Daiquiri and Margarita, they are not. The same goes for more modern creations. The Earl Grey MarTEAni is made with an egg white; the Cosmopolitan is not.

With Whiskey Sours, at least in the United States, there is little precedent for using egg whites prior to recent years. But today, in the best cocktail bars, it's hard to find a Whiskey Sour without an egg white. And I don't mean just a touch of egg white or egg white as part of a housemade sour mix; I mean a whole egg white.

Where did these bars and bartenders get this idea? Certainly not from the old cocktail books everyone was studying ten-fifteen-twenty years ago.

The outlier among English-language cocktail books of the past is *Cocktails and How to Mix Them* (1922) by Robert Vermeire, a London bartender. Vermeire, in the section on Sours, under the heading for Brandy Sour, wrote, "A few drops of white of egg improves all Sours." I suspect a lot of young bartenders of the aughts and early 2010s took that line to heart.

Some argue that the egg-white Whiskey Sour is actually a misnamed Boston Sour, a drink that appeared regularly in the Mr. Boston cocktail manuals.

American newspaper articles about egg-white Whiskey Sours being a revelation to customers began to appear around 2009–2011, in papers in Charlotte, Louisville, Los Angeles, and Madison.

My favorite theory about why Whiskey Sours came to be universally served as egg white cocktails comes from Bobby Heugel, owner of several bars in Houston. "I would bet that most 'first' cocktail bars in each respective city are somewhat similar," he told me in 2025 when I was working on a Whiskey Sour story for my newsletter *The Mix*. "Making the drink this way was not only a contrasting experience for guests who ordered Whiskey Sours, but it was also a visually theatrical contrast for any guests with eyesight. . . . The truth is that it was important for bars at that time to be iconoclastic and carve a space out in their local market, and cracking an egg into a drink everyone had heard of and thought they knew was very defining."

NEW YORK SOUR

This is a Whiskey Sour variation that has become popular in the last two decades. Floating the red wine on top can be tricky and takes practice.

2 ounces bourbon or rye	Combine all ingredients except the red wine in a cocktail shaker half filled with ice. Shake until chilled. Strain into a rocks glass filled with ice. Pour the red wine over the back of a barspoon, tilted and held close to the top of the cocktail so that the wine floats on the top of the cocktail.
1 ounce fresh lemon juice	
1 ounce Simple Syrup (see page 23)	
½ ounce red wine, for float	

1870
OLD FORESTER

Bourbon wasn't only originally stored and aged in barrels; it was also sold in them. If you wanted bourbon, you bought a barrel. If you wanted a drink or a bottle of that bourbon, you went to the local saloon or grocery store and it was dispensed from a barrel that the establishment had bought from the distiller, wholesaler, or other sort of middleman. Old Forester changed that in 1870 when it became the first bourbon to be sold exclusively in bottles. This made Old Forester an expensive bourbon, as bottling was a costly process; it would be decades before bottles could be mass produced.

Old Forester was the work of George Garvin Brown—the Brown of what would become the bourbon giant Brown-Forman (see page 94)—a former pharmaceutical salesman who sourced whiskey, bottled it, and sold it to doctors as reliable, untampered medicine (because of the sealed bottle). Doctors often prescribed whiskey as medicine and wanted a brand they could trust that had not been diluted or adulterated. This angle was so important that Brown actually named the brand after a doctor friend of his named Forrester. Remember how three of four dentists used to recommend Crest toothpaste? Well, this doctor recommended Brown's whiskey for his ailing patients.

After a few decades, Brown began distilling his own whiskey. Ever the salesman, when confronted with the growing temperance movement, he published a pamphlet in 1910 titled *The Holy Bible Repudiates Prohibition.* It was full of Bible passages that let teetotalers in the temperance movement know how wrong they were to call themselves Christians. It was a national bestseller.

The Brown family is still in charge of Brown-Forman. Campbell Brown, the great-great-grandson of George, was chairman until recently. (Forman was the accountant. So a company named after a salesman and an accountant sells a bourbon named after a doctor.)

Ironically, given its place in the company's history, Old Forester is kind of a wallflower brand at Brown-Forman. Its most famous product by far is Jack Daniel's, while its prestige whiskey is the newer Woodford Reserve. Aficionados, however, still know that the relatively unsung Old Forester—at least in its choicest expressions—is the distiller's best whiskey.

JERKWATER

SOTHER TEAGUE, AMOR Y AMARGO, NEW YORK CITY (2017)

This drink is Teague's American whiskey answer to the Cosmopolitan, the old term jerkwater—meaning a remote, insignificant backwoods burg—being the relative opposite of cosmopolitan.

- 1½ ounces Old Forester Rye
- 1 ounce curaçao
- ¾ ounce fresh lemon juice
- ¼ ounce Hibiscus Tea Syrup (recipe follows)
- Hibiscus flower or lemon twist, for garnish

Shake the liquid ingredients in a cocktail shaker filled with plenty of ice. Strain into a chilled coupe. Garnish with the hibiscus flower or lemon twist.

HIBISCUS TEA SYRUP MAKES ABOUT 2 CUPS

- ¼ cup dried hibiscus flowers
- 2 cups sugar
- 2 cups water

In a medium saucepan, bring the sugar and water to a simmer over medium heat and cook until sugar dissolves. Remove from the heat and add the hibiscus flowers. Allow to cool. Strain and discard the flowers. Store in an airtight container in the refrigerator for up to 1 month.

LINCOLN

**CHUCK CERANKOSKY,
BAR BANTAM, ROCHESTER, NEW YORK (2018)**

This cocktail, named after the Lincoln Tower—an undulating white skyscraper in Rochester that is home to Bar Bantam—is a Manhattan variant in the style of the "Neighborhood Cocktails" of the aughts, like the Greenpoint and the Red Hook. The Madeira, meanwhile, furnishes it with the character of a classic cocktail from the old days.

2 ounces Old Forester 86 proof Bourbon

½ ounce Cap Corse Rouge Quinquina

½ ounce H&H Full Rich Madeira

2 dashes of Fee Brothers Cardamom Bitters

Apple slice, for garnish

Combine the liquid ingredients in a mixing glass half filled with ice. Stir until chilled. Strain into a rocks glass filled with one large ice cube. Garnish with the apple slice.

1870
BUFFALO TRACE

The Buffalo Trace distillery is the Tiffany's of Kentucky whiskey. It's home to all the rare bourbon baubles that the wealthy and enthusiastic lust after. Pappy Van Winkle has been contract distilled here since 2002. They make the Weller line, which used to be affordable, but is no longer since folks were tuned in to the idea that it could act as a Pappy stand-in. The Antique Collection, an annual release of pricey whiskeys including George T. Stagg, Eagle Rare, Sazerac Rye, Thomas H. Handy, and William Larue Weller—to which Colonel E. H. Taylor Bottled-in-Bond Bourbon was added in 2025—comes out of here. Even its flagship brand, Buffalo Trace, has gone up in price and become harder to find over the past decade.

Like Tiffany's, Buffalo Trace does business out of a beautiful location and has a long history. Whiskey has been made since the 1700s on the property—located in a beautiful sloping valley along the Kentucky River near Frankfort that is probably the most picturesque in all of bourbondom. The property's modern history begins in 1870, when Edmund H. Taylor, Jr., acquired the land and started distilling. (A few years ago, Buffalo Trace uncovered Taylor's old fermentation rooms and started using them again. They call the space "Bourbon Pompeii" and that's not hyperbole.)

George T. Stagg was the next owner, the same Stagg whose name blesses Buffalo Trace's famous overproof bourbon (see page 174). In fact, unlike most Kentucky distilleries, the names on many Buffalo Trace whiskeys are of people who have an actual connection to the company. Blanton's, the trailblazing single-barrel bourbon, was named after Albert Blanton, who ran the distillery for decades in the early twentieth century. Elmer T. Lee, who succeeded Blanton, and who invented Blanton's, has his own namesake whiskey as well. As for Buffalo Trace bourbon, it's a relatively new product. The name comes from a story about Buffalo passing through the property centuries ago. You can thank the Sazerac Company, a New Orleans outfit that bought the property in 1992, for that corny story.

Buffalo Trace has an experimental bent. The distillery began toying around with different whiskey-making methods—various mash bills, a variety of barrel woods, different barrel toasts—in the early 1990s. Every possible variable was scrutinized. They were early actors in the single-barrel (see page 136), overproof, and white dog (see page 200) crazes. They're still at it, and fairly often they unveil one of their experiments as a limited release, with the first coming back in 2006. Dozens have been released and there are many thousands of experimental barrels sitting in Buffalo Trace rickhouses, waiting for their possible close-up.

Most recently, they have begun playing around with harnessing the aging process. In their Warehouse P is a second, smaller, temperature-controlled warehouse where the company stores barrels it doesn't plan to bottle for decades. Few rickhouses are climate controlled in bourbon country. But inside Warehouse P, it is always 45 degrees. The room functions like a large meat locker. The idea behind this deep freeze is to slow the aging process, which will theoretically retain more of the whiskey's fruit flavors while keeping at bay the unpleasant woody notes associated with over-aged bourbon. Some whiskeys won't see the light of day until 2068.

Think of it as another diamond necklace for the glass showcase of Buffalo Trace's future.

BROWN DERBY

HOLLYWOOD COCKTAILS,
AUTHOR UNNAMED (1933)

This drink was named in honor of the Brown Derby restaurant, a famous haunt of celebrities in mid-twentieth-century Los Angeles.

2 ounces Buffalo Trace bourbon	Combine the ingredients in a cocktail shaker half filled with ice and shake until chilled. Strain into a coupe.
1 ounce fresh grapefruit juice	
¾ ounce Honey Syrup (recipe follows)	

HONEY SYRUP MAKES ABOUT 1 CUP

1 cup honey	In a medium saucepan, bring the honey and water to a simmer over medium heat and cook until honey dissolves. Remove from the heat and allow to cool. Store in an airtight container in the refrigerator for up to 1 month.
½ cup water	

1882

OLD GRAND-DAD

Along with Old Overholt, Old Crow, and a few other "Olds," Old Grand-Dad is one of those ancient brands that people don't tend to give much thought to, because it has always been there, has never been a star, and probably will never disappear. We take it for granted and maybe look down on it a little. It doesn't help that its owner, Jim Beam, which acquired the brand when it bought National Distillers in 1987, doesn't give it much love. But true connoisseurs know that the bonded version (see page 84) is one of the great value bourbons around and that the overproof 114 is downright great.

Grand-Dad was a person. That person was nineteenth-century Kentucky distiller Basil Hayden, who also has another Beam whiskey named after him, making him perhaps the only person with two bourbon brands to his name. Basil's grandson was Raymond B. Hayden, who founded Old Grand-Dad's first distillery in 1882. Hayden died and left no heirs. Thereafter, the brand changed hands many times. It was the flagship whiskey of the Wathen distilling family for decades on either side of Prohibition.

Old Grand-Dad is a high-rye bourbon, something that's come back into vogue in the past quarter century. Also, its 114 proof, launched in 1981, was one of the first barrel-strength bourbons (see page 174). It predates Booker's and George T. Stagg. It's also among the cheapest.

STONE FENCE

The Stone Fence was a common rustic highball in the nineteenth century and a perfect fit for an old nineteenth-century brand like Old Grand-Dad.

1½ ounces Old Grand-Dad Bonded Bourbon	Add the liquid ingredients to a Collins glass filled with ice and stir to combine. Garnish as desired with a mint sprig, apple slice, or lemon twist.
2 ounces apple cider	
2 dashes of Angostura bitters	
2 dashes Peychaud's bitters	
Dash of orange bitters	
Mint sprig, apple slice, or lemon twist, for garnish	

1884
THE MANHATTAN

The Manhattan is more of a cocktail than that other major bourbon/rye delivery system, the Old-Fashioned (see page 80). I don't mean it is a better cocktail. I mean it is more like a cocktail—that is, a mixed drink. The Old-Fashioned is basically sweetened whiskey, an adult kiddie cocktail. Its ingredients beyond the whiskey—the sugar, the bitters, the garnish—are all accents. The Manhattan, however, must share its glass with a near-equal in terms of flavor: sweet vermouth.

Whiskey met vermouth in the 1870s, when the aromatized European wine became more available in the States. Vermouth, bartenders quickly found out, is a great collaborator, marrying well with almost any spirit. In short order, the world was made better by a raft of original vermouth cocktails that quickly became classics, including the Martinez, Martini, Rob Roy, and Manhattan.

The Manhattan predated the all-mighty Martini by a few years, first appearing in a cocktail book in 1884. Though stories vary, it is generally agreed that the drink was created in the New York borough with which it shares a name, though it probably took that name from the Manhattan Club, a private organization for Gotham's well-heeled and well-connected. Soon, it was the talk of the town and then, the nation.

Early recipes don't specify which whiskey went into the drink, but given its East Coast origins, rye would be the sensible guess. As the cocktail made its way unchallenged through the decades, bourbon became a more common choice, especially as rye began to die on the vine in the last years of the twentieth century. Many bars also made it with blended whiskey, which did not serve the drink well and robbed it of its once-fine reputation. But in the current century, the bartender soldiers of the cocktail revolution began to make the drink with rye again, thinking the choice more authentic.

On the coasts, the cocktail is commonly served "up"—that is, in a stemmed cocktail glass sans ice. But in the heartland, the Manhattan is often served on ice in a rocks glass, a style that emerged in the 1950s. In either case, it is the most important cocktail that still demands a cocktail cherry as its garnish. But that's okay. No drink asks for an olive except the Martini. At the top of the cocktail mountain, singularity connotes greatness.

DEATHBED MANHATTAN

ALLEN KATZ,
BROOKLYN, NEW YORK (2014)

This sturdy Manhattan comes from Allen Katz, a Brooklyn-based cocktail burgher who knows a thing or two about American whiskey, having made estimable rye at his New York Distilling Company for nearly two decades. This drink originally called for Ragtime Rye, which was the original product NYDC put out. But Jaywalk, their newer and even better expression, will do just as well. As for the name, you've probably guessed that this cocktail would be Allen's preferred final quaff.

- 2 ounces Jaywalk Rye
- ½ ounce Punt e Mes vermouth
- ½ ounce Martini & Rossi sweet vermouth
- 2 dashes of Angostura bitters
- Cocktail cherry, for garnish

Combine the liquid ingredients in a mixing glass half filled with ice and stir until chilled. Strain into a chilled coupe. Garnish with the cherry.

CLASSIC MANHATTAN

This is the standard formula for a traditional Manhattan.

2 ounces bourbon or rye

1 ounce sweet vermouth

2 dashes of Angostura bitters

Cocktail cherry, for garnish

Combine the liquid ingredients in a mixing glass half filled with ice and stir until chilled. Strain into a chilled coupe. Garnish with the cherry.

PERFECT MANHATTAN

"Perfect" in this context doesn't mean "better," but is rather the historical term for a cocktail that splits the difference between sweet and dry vermouth.

- 2 ounces bourbon or rye
- ½ ounce sweet vermouth
- ½ ounce dry vermouth
- 2 dashes of Angostura bitters
- Cocktail cherry, for garnish

Combine the liquid ingredients in a mixing glass half filled with ice and stir until chilled. Strain into a chilled coupe. Garnish with the cherry.

BOOTHBY

This cocktail was the luxe namesake drink of "Cocktail Bill" Boothby, the most prominent San Francisco bartender in the years leading up to Prohibition. It works best when made with only the finest whiskey and Champagne. So, save this one for special occasions.

- 2 ounces bourbon or rye
- 1 ounce sweet vermouth
- 2 dashes of Angostura bitters
- Cocktail cherry, for garnish
- 1 ounce Champagne

Combine the liquid ingredients in a mixing glass half filled with ice and stir until chilled. Strain into a chilled coupe. Garnish with the cherry. Top with Champagne.

BLACK MANHATTAN

TODD SMITH,
CORTEZ, SAN FRANCISCO (2005)

This is a Manhattan variation that has since gone on to be a modern classic, served at many bars.

2 ounces rye	Combine the liquid ingredients in a mixing glass half filled with ice and stir until chilled. Strain into a chilled coupe. Garnish with the cherry.
1 ounce Amaro Averna	
Dash of Angostura bitters	
Dash of Regan's orange bitters	
Cocktail cherry, for garnish	

1888
THE OLD-FASHIONED

Two cocktails have ferried more bourbon and rye into drinkers' mouths than any others: the Old-Fashioned and the Manhattan (see page 72). The Old-Fashioned is the older and simpler of the two. It is the direct descendent of the Whiskey Cocktail of the late eighteenth/early nineteenth century, which can be considered the original cocktail, so closely does it follow the spirits-sugar-bitters-water formula for that American invention. It's sweetened whiskey, basically, and it has sweetened millions of palates and lives since its creation.

The history of the Old-Fashioned (which you can discover in more detail in my earlier book *The Old-Fashioned*), falls into six basic chapters.

First came the original cocktail. And it was indeed the original cocktail, as defined by an 1806 article in the *Balance and Columbian Repository*: a mixture of spirit, sugar, bitters, and water. We recognize this today as the formula for the Old-Fashioned. When made with whiskey, the drink was called a Whiskey Cocktail back then and for many decades after.

Along about the 1860s came the Improved Whiskey Cocktail. This was a Whiskey Cocktail that the bartender "improved" by adding something extra, such as maraschino liqueur or Chartreuse. Some agreed the drink was improved this way. Others did not.

Which led to The Correction, around about the 1880s, and the advent of the adjectival addition "Old-Fashioned," as in "I'll have an Old-Fashioned Whiskey Cocktail." Emphasis on the Old-Fashioned. No fussing about, just the good old original Whiskey Cocktail. (The first published recipe for this version of a Whiskey Cocktail appeared in the book *Bartender's Manual* by Theophile Proulx, a Chicago bartender, in 1888.)

Then, in 1920, came Prohibition. And somewhere in the thirteen terrible years that followed came The Muddling. We don't know why, or when. But after Repeal, everyone was muddling orange and cherry in the bottom of their Old-Fashioneds, or plopping the fruit on top, or both. And unfruited Old-Fashioned was a thing of the past. Until . . .

The Return! Modern mixologists of the early twenty-first century struggled and wrestled with the Old-Fashioned until they finally realized that simplicity of form and technique was the way to restore this classic cocktail to classic status. Once again, the call was: spirit, sugar, bitters, water. No fruit. Only a twist, lemon or orange or both. The result? The Old-Fashioned became one of the most popular cocktails in the world again, for the first time in thirty years.

And, finally, came The Variations. Or, shall we say, Revenge of the Improved Whiskey Cocktail. Once again, for the past decade or so, we're being served Old-Fashioneds with a little bit of this and a little bit of that in them. Obscure amaros. Unusual bitters. Some don't even have whiskey as their base spirit. But the best Old-Fashioneds still have a whiskey foundation. And by whiskey, I mean bourbon or rye.

VALKYRIE OLD-FASHIONED

AARON POST,
VALKYRIE, TULSA, OKLAHOMA (2013)

The drink, the house Old-Fashioned at Valkyrie, the elder statesman of the Tulsa cocktail scene, is remarkable in that its base spirit splits the difference between bourbon and rye, thus giving the drinker the best of both worlds.

1 ounce Elijah Craig bourbon

1 ounce Rittenhouse rye

¼ ounce Demerara Syrup (see page 43)

2 dashes of Angostura bitters

Orange twist, for garnish

Combine all liquid ingredients in a mixing glass half filled with ice. Stir until chilled. Strain over one large ice cube in a rocks glass. Garnish with the orange twist.

1897
BOTTLED-IN-BOND

For a spirit born of rebellion and overly fond of adopting an independent persona, the bourbon industry and the federal government have joined forces on several important occasions, usually to the bourbon makers' great benefit. The first such occasion was in 1897, when the Bottled-in-Bond Act was passed by Congress. This piece of legislation, which is still doing its good work today, protected distillers and consumers from bad whiskey.

See, back then the whiskey world was replete with opportunistic and untrustworthy middlemen. Distillers sold their liquid in barrels to rectifiers, wholesalers, and shopkeepers. And what happened to the juice after that, nobody knew. The whiskey could be watered down, adulterated, and generally messed around with. Then it was bottled under the distiller's name and sold, giving the distiller a bad name if the end product was indeed bad. Think of the unscrupulous bar owners who still exist today who pour out half of a bottle, fill it with water, and sell you shots of supposedly original product at regular prices. That was happening all the time, but on a larger scale.

Assurances were needed to bolster consumer confidence. And pledges and promises from the distiller were not going to cut it. Some sort of official stamp of approval from the highest seats of government was needed. At the urging of the distiller Edmund Taylor, Kentucky senator and former Confederate colonel J. C. S. Blackburn—whose name sounded like a bourbon brand—sponsored the bill.

The bill accomplished many things all at once. Bonded whiskey, in order to be deemed so, has to be produced by a single distiller at a single distillery during a single season, aged for at least four years, and bottled at 100 proof. These were all good things. To mark a bottle as having passed this test, its neck was affixed with a green stamp, which soon became a familiar sight for consumers. Bonded whiskey quickly developed a reputation as the good stuff. In the 1956 Douglas Sirk film *Written on the Wind,* when a bartender chastises millionaire oilman Robert Stack for liking corn whiskey, he says, were their positions switched, he would drink, "bonded bourbon and nothing but."

During bourbon's doldrum days in the late twentieth century, when it was competing with clear spirits for the dollars of drinkers who sought a lighter touch, not as many Kentucky distillers bothered with going bonded but instead brought their alcohol levels down to as low at 40 percent. However, that all changed in the early years of the current century. Bourbon enthusiasts and bartenders wanted whiskey with more punch and flavor. Bottled-in-Bond bottlings came into vogue again. Whiskeys that clocked in

at 100 proof and four years of age, like Rittenhouse rye, became the darlings of the industry. Even today, when a major name like Bulleit or George Dickel comes out with a bonded whiskey, it's big news.

THE BOURBON BRAMBLE

GARY CRUNKLETON,
THE CRUNKLETON, CHAPEL HILL,
NORTH CAROLINA (2010)

Crunkleton is a prominent North Carolina bartender who owns multiple bars across the state. This drink is his bourbon spin on London cocktail revival godfather Dick Bradsell's modern classic cocktail, the Bramble, which is made with gin and blackberries. (Use quality cocktail cherries for this recipe, not the common maraschino cherries found in the supermarket.)

- 2 ounces bonded bourbon
- 10 Marasca cherries and their syrup
- 1 ounce fresh lemon juice
- ¼ ounce maraschino liqueur
- 1 barspoon Simple Syrup (see page 23)
- Lemon wheel, for garnish

Place the liquid ingredients into a cocktail shaker and muddle the fruit. Muddle the cherries a lot because they serve as a sweetener. Add ice and shake until chilled. Using a Hawthorne strainer, strain the cocktail into a julep cup filled with crushed ice, allowing the muddled cherries to pass through. The cherries will serve as part of the garnish. Garnish with the lemon wheel.

1899

THE SAZERAC

This cocktail is about rye. Not Cognac. Because Cognac was never the base spirit of the Sazerac. Not even a mix of Cognac and rye. No Cognac ever.

Popular myths are sticky things. In the early years of the cocktail renaissance, it became conventional thinking by bartenders and cocktail writers to insist that the Sazerac—a New Orleans cocktail that many people erroneously believed to be the oldest original cocktail—was originally made with Cognac. They believed this primarily because of the bad reporting of a man named Stanley Clisby Arthur, who, in 1937, wrote a book called *Famous New Orleans Drinks & How to Mix 'Em*. He said the name of the drink was in reference to the Cognac brand Sazerac de Forge & Fils—from which the famous Sazerac Coffee House, founded in the 1850s, also took its name—and that the recipe was changed from brandy to rye "to please the tastes of Americans who preferred 'red likker.'" More contemporary authorities insisted the change came about because of the outbreak of phylloxera in France, which decimated the grapevines Cognac producers depended upon to make their spirit. Faced with a Cognac vacuum, Americans turned to rye whiskey.

None of this is true, as proven by cocktail historian David Wondrich in the updated edition of his cocktail history, *Imbibe!*, and later in the *Oxford Companion to Spirits and Cocktails*. No mention of a Sazerac cocktail appears in any paper before 1899 and, then and afterwards, it was always a rye cocktail. As for the name, it came from the series of bars that bore that name rather than the brand of Cognac.

Regardless of this easily acquirable edification, there are still plenty of bars that say they serve the Sazerac "correctly" with Cognac. But most bars use rye, including most in New Orleans. The Sazerac was once a nationally known and enjoyed cocktail. But after Prohibition, it was primarily the cocktail-hour habit of New Orleanians. There was no mystery to this. Two of the necessary ingredients—Peychaud's bitters and Herbsaint (the original absinthe had been banned by that point)—were made in New Orleans.

New Orleans was where I discovered it in 2006. But, soon after, it began to pop up at various quality cocktail bars in New York, usually served by bartenders who had been to New Orleans to attend the Tales of the Cocktail convention. The Sazerac, along with the Manhattan and Old-Fashioned, was one of the main reasons why bartenders and drinkers began to cry out for the return of quality rye whiskey. They wanted to make the drink the way it once had been. Rye whiskey and the Sazerac cocktail thus became great friends.

THE SAZERAC

2 ounces rye	Fill one rocks glass with ice. While it is chilling, combine the rye, simple syrup, and bitters in a mixing glass half filled with ice and stir until chilled. Empty the first glass of its ice and rinse the inside of the glass with a barspoon of absinthe. Pour out the excess absinthe. Strain the rye mixture into the waiting glass. Express the lemon twist over the drink and discard.
Barspoon of Simple Syrup (see page 23)	
4 or 5 dashes of Peychaud's bitters	
Absinthe	
Lemon twist, for garnish	

1909

THE TAFT DECISION

As Presidents go, William Howard Taft stands out. He only served four years (1909–1913), but it was hard to miss him, with his enormous girth and handlebar moustache—he was, in fact, the last President to sport facial hair. He is also the only U.S. President to subsequently serve as Chief Justice of the Supreme Court, and the first buried in Arlington National Cemetery. To boozehounds, however, he will forever be the author of the Taft Decision, which helped codify bourbon.

The Pure Food and Drug Act had passed in 1906, protecting Americans from unscrupulous producers. Whiskey distillers liked it. Whiskey rectifiers and blenders, not so much. They were made to label their products with descriptors like "imitation whiskey," which doesn't sound too yummy. The issue wouldn't go away, so once Taft was in office, he set about to determine the true meaning of the word "whiskey" as it applied to American production of hard spirits. (Taft arguably had a family interest. His ancestor, Samuel Taft, a Revolutionary War soldier, owned a tavern in Massachusetts that once hosted Washington.)

Taft took a year coming to his conclusions, which he revealed just after Christmas in 1909. *The New York Times* reported the results of Taft's inquiry:

"In brief the regulations declare that all unmixed spirits distilled from grain, prepared in the customary ways, are entitled to the name 'whiskey,' without qualifications. Blended whiskey must be labeled as such. The term 'whiskey,' however, is restricted to distillates from grain, and, under the regulations, distillates from other substances, if labeled 'whiskey,' are misbranded, and the person guilty of misbranding may be prosecuted. All descriptive matter of the labels must be accurate, and the word denoting the grade or kind of whiskey must be in type as large as the word 'whiskey' itself."

The Washington Post, meanwhile, didn't bury the lede: "Whiskey is whiskey, President Taft has at last decided," the paper wrote.

The decision defined what was meant by "straight whiskey." That term was already in use, but Taft said it should mean only pure grain whiskey cut with only water and unadulterated by other things.

Those who didn't like the new law bitterly referred to the spirits that abided by its definitions as "Taft Whiskey."

One hundred years later, Taft was still important enough to the history of the whiskey industry that the Kentucky Distillers' Association placed him in the Kentucky Bourbon Hall of Fame. He is the only president so honored.

TAFT COCKTAIL

**CHRIS HANNAH,
NEW ORLEANS (2025)**

To celebrate the Taft Decision, an unknown person in New Orleans created the Taft Cocktail in his honor. A 1909 article had it that "half" the town was drinking it. But it did not say how the drink was made. It said just, "The Taft cocktail is built after the manner of the Creole cocktail but it has some trimmings. A mixture of lemon and lime juice is prepared; the rim of the glass, after being iced, is dipped in this mixture and frosted in confectioner's sugar." I handed over this information to noted New Orleans bartender Chris Hannah and he came up with the below recipe, inspired by the Creole cocktail recipe of famous New Orleans bartender Henry Ramos, who was doing business at the time of the Taft Decision. Hannah told me, "Everything in this recipe was in Ramos's Creole except the citrus and sugar rim."

- 1½ ounces rye or bourbon (Hannah prefers bourbon)
- 3/5 ounce 50/50 mixture of lemon and lime juice
- ½ ounce curaçao, preferably Ferrand
- ¼ ounce absinthe, preferably Lucid
- 4 dashes of Angostura bitters
- 5 dashes of Peychaud's bitters
- Confectioners' sugar, for rim
- Lemon or lime wedge, for rim

Moisten the rim of a small wine glass with a lemon or lime wedge and dip in confectioners' sugar. In a cocktail shaker, combine the remaining ingredients and shake until chilled. Strain into the prepared glass. Serve without a garnish.

1920–1933
PROHIBITION

Looked on as a whole, the twentieth century was not good for bourbon and rye. Distillers had as many bad years as they had good. It was basically one long series of potentially industry-endangering knock-out punches. There was World War II, when all the distilleries' energies had to be channeled toward the war effort. There was the period from the 1970s on, when public tastes changed and nobody seemed to want whiskey at all. But no TKO was as bad as the first one, Prohibition, a thirteen-year failed experiment to see if an entire nation could get by without access to strong spirits. (Couldn't; didn't want to.)

Obviously, the American whiskey industry is still with us today, so the distillers found some way to survive. Except they didn't. Most companies died. Only a few toughed it out and made it to the other side, and they did this by way of a few loopholes in the Volstead Act.

In order to stay alive, whiskey went back to its quasi-medical roots. A handful of lucky distilleries were able to continue producing whiskey for medicinal purposes, courtesy of Title II, Section 6 of the Volstead Act. And, wouldn't you know it, a lot of people turned out to be feeling poorly, and a lot of doctors agreed that said people needed a dose of brown, distilled, aged medicine.

The six companies that graduated from recreational to medicinal included A. Ph. Stitzel Distillery, Schenley Distillers Corporation, Brown-Forman, Frankfort Distilleries, American Medicinal Spirits Company, and James Thompson and Brother. All the "medicine" had to be drawn from bonded warehouses. When they ran out, the government passed a law to allow them to make more whiskey. Some of these companies gobbled up the smaller distilleries that were going out of business, acquiring their whiskey brands in the process. In short, Prohibition was great for capitalistic conglomeration and terrible for individualistic innovation. The rich got richer and the poor got poorer.

As for the public that didn't have the sniffles or couldn't connect with a cooperative physician, they got their booze at one of the thousands of speakeasies that popped up overnight. The whiskey served at these joints was usually Canadian or Scotch, bootlegged into the country, much of it watered down or otherwise adulterated.

The effects of Prohibition are still felt today, nearly a century later, but more in aesthetic terms. The craft cocktail bars that kicked off the cocktail renaissance in America called themselves speakeasies. But these speakeasies served real bourbon and rye, and often the best available.

AUTUMN LEAVES

FRANCIS SCHOTT, STAGE LEFT STEAK, NEW BRUNSWICK, NEW JERSEY (2011)

With his partner Mark Pascal, Schott has run the best craft cocktail program in New Jersey for more than thirty years. For more than a decade, the two used to come up with a new cocktail every week. "I love a hot toddy, but hot drinks don't work in every application and they really don't work when the mercury is high," Schott told me. "This is basically the warm autumnal flavors of the hot toddy but cold."

- 2 ounces Old Forester 1920 Prohibition Style Bourbon
- 1/4 ounce St. Elizabeth Allspice Dram
- 1/2 ounce fresh lemon juice
- 1/2 ounce Honey Syrup (see page 69)
- 1/4 ounce Cinnamon Syrup (recipe follows)
- 1/4 ounce Vanilla Syrup (recipe follows)
- Grated cinnamon and nutmeg, for garnish

Combine the liquid ingredients in a mixing glass half filled with ice and shake until chilled. Strain into a sour glass over one large ice cube. Garnish with the grated cinnamon and nutmeg.

CINNAMON SYRUP MAKES ABOUT 1 CUP

1 cup sugar

1 cup water

1 cinnamon stick

In a medium saucepan, bring the sugar and water to a simmer over medium heat and cook until sugar dissolves. Add the cinnamon stick and steep for 15 minutes. Remove from the heat and allow to cool. Remove and discard the cinnamon stick and store the syrup in an airtight container in the refrigerator for up to 1 month.

VANILLA SYRUP MAKES ABOUT 1 CUP

1 cup sugar

1 cup water

1 vanilla bean

In a medium saucepan, bring the sugar and water to a simmer over medium heat and cook until sugar dissolves. Add the vanilla bean and steep for 15 minutes. Remove from the heat and allow to cool. Remove and discard the vanilla bean and store the syrup in an airtight container in the refrigerator for up to 1 month.

1934
JIM BEAM

Like the Snopes family in the novels of William Faulkner, you can't dig too far into the history of bourbon without running into a Beam. You'd think all the Beams would be working down on the Jim Beam place in Clermont, Kentucky. You'd be wrong. There have been too many of them, so you can find a Beam in the history of almost every major American whiskey distillery of the twentieth century and beyond.

A few examples. Parker Beam was the master distiller at Heaven Hill (see page 104) for decades; his name is on their limited release series Parker's Reserve. Park Beam was the distiller at Willett (see page 108) for many years following Repeal. Roy Beam was distiller at Frankfort Distillery, which made Four Roses (see page 176). Everett Beam was master distiller at Michter's in Pennsylvania (see page 152) before that brand's recently reincarnation as a Kentucky whiskey.

The Beam in Jim Beam Bourbon is James Beauregard Beam, the great-grandson of Johannes Jakob Boehm, the distiller who kicked off the whole family tradition back in the late eighteenth century. (His middle name, Beauregard, commemorates a Confederate general.) It's weird to think that the Jim Beam distillery, the most successful bourbon producer in the world, was once an upstart, but it was in a way. James B. Beam distilled his brand, Old Tub, for decades until his business became one of the casualties of Prohibition. (No government contract for medicinal whiskey for Beam.) But unlike most of his colleagues, he decided to try again in 1934, when Prohibition was history, building a new distillery in Clermont, Kentucky, with the help of some Chicago investors. Soon enough, the world became familiar with Jim Beam White Label, the company's benchmark product and still one of the most famous and ubiquitous whiskey brands today. Beam died in 1944, just a decade after founding the new operation, and handed the reins off to his son, Jeremiah.

Today, Jim Beam is no longer run by people named Beam. It's run by people named Noe. T. Jeremiah Beam didn't have any children, so he brought in his nephew, Frederick "Booker" Noe II, son of Margaret Beam Noe. Noe *did* have children, and grandchildren, and they all took his name, so we have Frederick Booker Noe III, who goes by Fred, and Frederick Booker Noe IV, who goes by Freddie. Under the Noes, the Beam brand diversified in many ways. Booker came up with a single-barrel, barrel-strength bourbon called, aptly, Booker's, in 1988, a time when no one was doing such things. It remains a release prized by enthusiasts today.

The Knob Creek line, introduced in 1992, boosted Beam's reputation for quality whiskey. Baker's and Basil Hayden's joined Knob Creak under the new "small-batch" umbrella. Beam also acquired other old brands, like the fancy premium Maker's Mark, populist mainstay Old Grand-Dad, and the old standby Pennsylvania rye Old Overholt. Between Overholt and Jim Beam Rye, Beam did more than anyone to keep the rye flame burning for decades. Beam also became an early innovator in the flavored whiskey category when it released the cherry-flavored Red Stag. In this way, even if you're not a Jim Beam fan, every bourbon and rye lover is probably enjoying a Beam product of some kind on the regular.

In 2014, Beam—which has been owned by various companies over the years—was bought by the Japanese giant Suntory. (Oddly, the Beam family has not owned the company since the 1940s.) The conglomerate was known as Beam Suntory until 2024, when it became Suntory Global Spirits.

RED SQUIRREL OLD-FASHIONED

BRIAN BARTELS,
SETTLE DOWN TAVERN,
MADISON, WISCONSIN (2020)

Bartels, a Wisconsin native who owns several bars in Madison but worked for a decade in Manhattan bars, created this drink as a halfway point between the two drinking cultures. The Korbel and maple syrup are pure Wisconsin; the Campari and bourbon are from the cosmopolitan East Coast.

- ¾ ounce Jim Beam bourbon
- ¾ ounce Korbel brandy
- ½ ounce Campari
- ¼ ounce maple syrup
- 3 dashes of Peychaud's bitters
- 2 dashes of Angostura bitters
- Orange twist, for garnish

Combine the liquid ingredients in a mixing glass half filled with ice and stir until chilled. Serve in an Old-Fashioned glass filled with ice. Garnish with the orange twist.

THE RUTHERFORD

MURRAY STENSON,
ZIG ZAG CAFÉ, SEATTLE

Stenson was the dean of Seattle cocktail bartenders until his death in 2023. A more complex version of a Black Manhattan, the Rutherford makes for a perfect nightcap.

- 1½ ounces Knob Creek bourbon
- ½ ounce Amaro Meletti
- ½ ounce Amaro Nardini
- ¼ ounce Bénédictine
- Orange twist, for garnish

Combine the liquid ingredients in a mixing glass half filled with ice. Stir until chilled and strain into a chilled cocktail glass. Garnish with the orange twist.

SILK STOCKINGS #1

BRYANT'S COCKTAIL LOUNGE, MILWAUKEE, WISCONSIN (CIRCA 1940s)

Bryant's is the oldest cocktail bar in Milwaukee, dating from the 1930s. It remains frozen in time, with a retro décor and a secret rolodex of house cocktails that goes back decades. Interviewed for this book, owner John Dye told me, "The cocktail was named after the streetcar that would bring well-dressed, affluent women to Mitchell Street to shop. They wore silk stockings and apparently there were so many that the streetcar was nicknamed the silk-stocking line." Please note that this drink will not taste exactly like it does at Bryant's, because the bar has a few secret techniques that it will not divulge. But it will be close.

1 ounce Jim Beam bourbon

1 ounce dry vermouth

1 ounce Mandarine Napoleon liqueur

1 ounce fresh lemon juice

1 ounce orange juice

½ ounce Simple Syrup (see page 23)

Orange twist, for garnish

Combine the liquid ingredients in a cocktail shaker half filled with ice and shake vigorously until chilled. Serve in a highball glass over ice. Garnish with the orange twist.

1935
HEAVEN HILL

Heaven Hill is every cocktail bartender's favorite Kentucky distillery. It's an unlikely choice. Heaven Hill isn't the oldest bourbon or rye producer; it was founded in 1935. It's not run by a marquee Kentucky whiskey family, like Jim Beam (the Noes) and Wild Turkey (the Russells). It was founded by five brothers named Shapira, who owned a string of department stores in Louisville. The Shapiras still own and run it, making it that rare thing: a large family-owned Kentucky whiskey distillery.

Heaven Hill is a bartender darling because it quietly makes all the value bourbons and ryes that bartenders like to mix with, including Elijah Craig, Evan Williams, Rittenhouse, Henry McKenna, Mellow Corn, and Bernheim. None of these workhorse whiskeys were celebrated before the mixology movement came along, but they are all venerated today, particularly Elijah Craig, which used to be a twelve-year-old bourbon that sold for peanuts. (Today, if you have a spare $1,500, you can buy a bottle of Elijah Craig 23-year-old.)

There's a little touch of old Kentucky heritage in the HH story. When it was founded, Joseph Beam, Jim Beam's first cousin, was the master distiller. The Shapira brothers, who were businessmen at heart and knew nothing of the craft of distilling whiskey, kept him in that position when they assumed full control of the firm. Many other Beams followed Joe in the master distiller role, including Parker Beam, who began working at Heaven Hill in 1960 and whose name graces Parker's Heritage Selection, the well-regarded series of limited experimental bottlings.

Heaven Hill's production plant suffered a massive fire in 1996 that consumed ninety thousand barrels of whiskey. Another distillery would have thrown in the towel, but, with a little help from its distillery friends, who let them use their facilities for a while, Heaven Hill actually emerged from the crisis stronger. In 1999, it bought the Bernheim distillery in downtown Louisville from the international liquor giant Diageo. Since then, the company has only grown.

Heaven Hill has a seemingly paradoxical personality. On the one hand, it is an industrial booze factory, pumping out more juice than everyone but Jim Beam. I visited the Louisville distillery once. The stills are multi-story mammoths and there is no romance to the place. They also have a knack for hokum, naming their brands after Kentucky frontiermen (Craig, Williams, etc.) with whose history they have nothing to do.

On the other hand, Heaven Hill benefits greatly, image-wise, from still being a family business, one that seems to value the small alongside the big (Evan Williams is the second biggest-selling bourbon in the world). Heaven Hill is well diversified; it makes 1,400 brands of spirits. That's a lot of brands. But Heaven Hill also produces the well-respected leading brands in less-trafficked American whiskey categories like wheat whiskey (Bernheim) and corn whiskey (Mellow Corn). It put money into creating a better version of Dubonnet, which it also owns, and an improved Christian Brothers brandy called Sacred Bond (now sadly discontinued). And it revived the name of Pikesville, a once-famous Pennsylvania rye. It aims high, low, and everywhere in between. Just like good businessmen would.

CHAI TOWN

**LYNN HOUSE,
CHICAGO (2013)**

Lynn House is not only a Chicago bartender with many years of experience, but also the brand ambassador for Heaven Hill. Who better to provide a recipe for this chapter? This cocktail draws its inspiration from House's favorite classic cocktail, the Manhattan. "In lieu of sweet vermouth, I opted for PAMA, because I love the way the tannins in pomegranates compliment the spice of the rye," House told me. "I enjoy using teas in my cocktails. It is an easy way to incorporate complex and unique flavors." The name is a tongue-in-cheek reference to House's hometown, often referred to as Chi Town.

1½ ounces Rittenhouse rye

¾ ounces PAMA pomegranate liqueur

1 ounce strong brewed chai (steep 2 bags chai to every 8 ounces hot water for a concentrated flavor)

¼ ounce Simple Syrup (see page 23)

3 dashes of Fee Brothers Old Fashion Bitters

Orange twist, for garnish

Combine all ingredients except the bitters in a mixing glass half filled with ice and stir until chilled. Add bitters and stir three more times. This will incorporate the flavor of the bitters without destroying its aromatics. Strain into a chilled cocktail glass. Express the orange twist over the cocktail and then place it on the rim of the glass.

1936

WILLETT

One day in the aughts, I took a seat at the bar at Jack Rose Dining Saloon, the whiskey destination in D.C., and began perusing the whiskey menu. Choosing a whiskey is tough at Jack Rose; they stock thousands of different kinds, both extant and extinct. Toward the end of the bourbon section of the menu, there was a heading for Willett. There were an awful lot of different Willett bottlings, dwarfing the number of choices of nearby Wild Turkey. I looked up and found the Willett section on Jack Rose's shelves behind the bar. The bottles were unusual. They were long and tall. The name Willett was in script, above a large family crest that looked more in keeping with Scotch than bourbon. Each label was remarkably transparent in its information, at a time when that wasn't usually the case. The label told you exactly how old the liquid was and which barrel it had come from. Lined up next to each other, the bottles looked more like labeled laboratory experiments than commercial whiskey products. And many of the selections were rye. That, too, was unusual.

I tried two of them. They were remarkable, some of the best ryes I'd ever had. Back home, I began looking for Willett. When I could locate it, it was usually on the highest shelf of the liquor store or kept behind glass. There were typically just one or two bottles available. I bought it when I could afford it. No bottle ever let me down.

Willett's an odd duck. It's an old name with a new sheen. The distillery in Bardstown goes back to 1936, when Thompson Willett, the scion of a family with long whiskey-making roots, opened it. But most people outside Kentucky had never heard of it until those special, antique-looking, one-off bottles started appearing at their local liquor stores and whiskey bars in the aughts. Willett also provided juice for various private bottlings. The most famous of these, and certainly the coolest, was Red Hook Rye, the now legendary whiskey that was sold by LeNell's, a boutique spirits shop that lived for a while in Red Hook, Brooklyn.

It's still run and owned by the Willett family, though the branch that operates it today is named Kulsveen; Even Kulsveen, a Norwegian immigrant, married into the Willett family and got the Bardstown property humming again after decades of dormancy, and his son Drew Kulsveen is the face of the brand today. (Even died in 2025.)

Willett makes some of its own whiskey today, but from the 1990s on it mainly sourced aged whiskey from around Kentucky. Starting in 2008, it bottled it under its own Willett Pot Still Reserve (in a bottle shaped like a pot still) and Willett Family Reserve labels.

1938

MINT JULEP

Ask any bartender in the United States about the Mint Julep and they will tell you it is a cocktail they can't give away except on one calendar date—Kentucky Derby day—when they sell them hand over fist.

Perhaps due to this lopsided situation, most bars are not prepared to make a Mint Julep on any old day, even if the demand arises. The drink, after all, requires ingredients and equipment that are not exactly run of the mill. These include fresh mint (and lots of it), crushed ice (and lots of it), and a julep cup—a metal tin, usually silver, that is the cocktail's traditional vessel.

The Silver Dollar, which sadly closed in November 2025, was not such a bar. The sprawling modern saloon, once located in the Clifton neighborhood of Louisville, was prepared to fill an order for a Mint Julep any day of the year. To former owner Brian Downing, who was once a bartender at the bar, this was a point of pride.

The Mint Julep is one of the oldest American mixed drinks on record. Accounts of it first appeared in the 1700s in Virginia, where it was a thirst-quencher of the upper classes. By the early twentieth century, it had drifted west to Kentucky and taken on that state's signature spirit, bourbon, as its base. In 1938, the Kentucky Derby hitched its wagon to the drink, making it the race's official cocktail. It has remained closely linked to the Derby ever since.

The Mint Julep famously calls for crushed ice. In recent years, retro-minded mixologists have met this need in a decidedly old-school fashion, by crushing ice chunks in a white canvas bag—known as a Lewis bag—with a large wooden mallet. It makes for quite a show and renders an already theatrical cocktail even more so. "That first sip is going to be intense," Downing told me for a 2025 article in *The Wall Street Journal*. "It is going to be mint, sugar, and bourbon exclusively." After that, however, the cocktail will swiftly dilute. "That drink is going to evolve over time. That cocktail is going to become more and more refreshing."

"It's kind of a joke, the idea that when bartenders don't want to make something, they say, 'We're all out of mint.' That's not hospitable," said Downing. "Give the people what they want. Mint is available all year round. This is the capital of American whiskey. To not give them that opportunity to have a julep on their visit to Kentucky—that is a challenge we have to rise to."

MINT JULEP

SILVER DOLLAR, LOUISVILLE, KENTUCKY (2024)

This was the house Mint Julep at Silver Dollar, where they were ready to make the cocktail any day of the year.

- 2 ounces Four Roses Small Batch Select (or a bourbon of the same high proof, which is 104)
- ½ ounce Strong Demerara Syrup (recipe follows)
- 6 or 7 fresh mint leaves
- More fresh mint leaves, for garnish

Muddle the mint leaves in the bottom of a julep cup. Discard mint. Add syrup and bourbon and briefly stir. Add enough crushed or shaved ice to create a dome that rises about an inch or two above the lip of the cup. Garnish with a generous cluster of good-sized mint leaves. Slip a straw (preferably metal) alongside the mint.

STRONG DEMERARA SYRUP

MAKES ABOUT ½ CUP

- 1 cup Demerara sugar
- ½ cup water

In a medium saucepan, bring the sugar and water to a simmer over medium heat and cook until sugar dissolves. Remove from the heat and allow to cool. Store in an airtight container in the refrigerator for up to 1 month.

JULIAN ANDERSON'S MINT JULEP

JULIAN ANDERSON,
***JULIAN'S RECIPES*, HELENA, MONTANA (1919)**

Anderson, who tended bar at the private Montana Club in Helena for sixty years beginning in the 1890s, was only the second Black bartender in the United States to publish a cocktail book, 1919's *Julian's Recipes*. He was most famous for his Mint Julep. "Rightfully recognized for fame are Julian's mint juleps, which have been discussed from Havana to Toronto, from San Francisco to Boston," wrote Ann McCabe in the *Independent Record* in 1943. Part of the secret to the Juleps' success was in the mint, which came from Anderson's own backyard garden, born of cuttings from his childhood home in Bowling Green, Virginia. "Many claim it is 'the mother of all mint in Helena,'" wrote McCabe. The recipe included in his manual testifies to the considerable time Anderson spent perfecting the drink.

- 1½ ounces brandy
- 1½ ounces bourbon
- 1½ ounces water
- 1 teaspoon Jamaican rum
- 4 sugar cubes
- 5 sprigs of mint
- More mint, for garnish
- Cherries, orange slice, and sliced pineapple, for garnish
- Powdered sugar, for garnish

Combine the sugar cubes, spirits, mint, and water in a mixing glass and stir until the sugar is dissolved. Fill a glass or julep cup with crushed ice. Strain mixture from mixing glass into ice-filled glass. Add more crushed ice if necessary. Stick a bunch of mint in the center of the glass. Garnish the drink with fruit. Sprinkle the surface with powdered sugar.

1942
WILD TURKEY

Today, the Wild Turkey distillery is associated with one name: Russell. Jimmy Russell, a local man who has worked there since 1954, became master distiller in 1967; Eddie Russell, his son, who joined his father in 1981 and succeeded him as head distiller; and Bruce Russell, Eddie's son.

The Russell name became known outside Kentucky when Wild Turkey released the Russell's Reserve line. Prior to that, you'd be forgiven for thinking the top man at Wild Turkey was some guy named Austin Nichols. That's because the name was on every bottle. But Austin, Nichols & Co. (Robert F. Austin and James E. Nichols) was actually a huge grocery wholesaler based in Brooklyn that contracted the distillery to make the private-label whiskey it sold. It was actually an Austin Nichols exec, Thomas F. McCarthy, who was with the firm for fifty years, who gave the whiskey its lasting name.

Eventually Austin, Nichols & Co. bought the distillery outright in 1971. It wasn't until then that the distillery was actually named Wild Turkey. It was a confusing label, the bourbon world's version of Ruth's Chris Steak House. Today, it is owned by Campari Gruppo, the Italian liquor conglomerate, making for the most unlikely Odd Couple business partnership in the bourbon world. As for Austin, Nichols & Co., the one-time largest grocery wholesaler in the world, a business behemoth, it is completely forgotten. (Go to Wikipedia and type in Austin, Nichols, and you'll be led to the Wild Turkey page.)

Wild Turkey has an odd, contradictory reputation. The bourbon world, for all its talk of tradition, is very changeable. When business was bad in the late twentieth century, many distilleries chased after the clear-spirit drinkers, lowering their ABVs and producing clear versions of whiskey. When the boom times of the last twenty years arrived, those same distilleries reacted to a willing customer base by throwing every line extension they could think of at the wall. Meanwhile, the Russells stuck to their guns. Their signature 101 proof bourbon remained the same through good times and bad. That is something of a miracle, a mark of the most admirable type of business stubbornness. Wild Turkey also continued making rye when most everybody had given it up for dead. When rye came roaring back, Wild Turkey looked prescient.

This stay-the-course approach had a split effect. On one side, people thought of Wild Turkey as a fuddy-duddy distillery, the whiskey your grandfather drank, unsexy, unchanging. Among enthusiasts and critics, however, Wild Turkey was seen as a standard bearer, the rare bottle on the shelf you could always count on.

SCOFFLAW

BILL BOOTHBY,
WORLD DRINKS AND HOW TO MIX THEM,
SAN FRANCISCO, 1934 EDITION,
ADAPTED BY THAD VOGLER

Bill Boothby was the preeminent cocktail bartender in the San Francisco of the late nineteenth and early twentieth centuries. Thad Vogler is one of the leading San Francisco cocktail bar owners of the early twenty-first century. Two Bay Area legends join forces in this recipe. Vogler adjusted the recipe by increasing the amount of whiskey, lemon juice, and grenadine and improving the choice of vermouth.

1½ ounces Wild Turkey 101 bourbon

½ ounce fresh lemon juice

½ ounce grenadine

½ ounce Bordiga Extra Dry vermouth

Combine the ingredients in a cocktail shaker half filled with ice. Shake until chilled and double strain into a chilled coupe or Nick & Nora glass.

THE GINGERMAN

KURT HERNON,
SPEAK OF THE DEVIL, LORAIN, OHIO (2017)

This cocktail is named after J. P. Donleavy's wild, post–World War II novel *The Ginger Man* (1955), one of the drinkingest books ever written. "The Gingerman cocktail was born of the extreme cold weather that gripped most of the United States when we opened Speak of the Devil in late 2017, early 2018," Hernon told me. "The bitter cold inspired us to reach for some high proof bourbon to thaw our souls out. Thus, the Gingerman became our very first ever original cocktail."

2 ounces Wild Turkey 101 bourbon

¾ ounce Domaine de Canton ginger liqueur

¾ ounce fresh lemon juice

½ ounce Cinnamon Syrup (see page 96)

2 dashes of Angostura bitters

Lemon twist, for garnish

Combine the liquid ingredients in a cocktail shaker half filled with ice. Shake until chilled. Strain into a rocks glass over one large ice cube. Garnish with the lemon twist.

1950s

FRANK SINATRA

Certain celebrities are of such a magnitude, and command such a huge fanbase, that they can have an outsized impact on the booze world by the smallest gesture. Ernest Hemingway could make a bar an internationally famous tourist destination simply by visiting it or writing a single line extolling it. Frank Sinatra also had this sort of pull. Like Hemingway, he seemed to be everywhere all at once. The number of restaurants that boast of a Sinatra booth runs into the dozens. And he and his Rat Pack buddies single-handedly made the Rusty Nail a popular cocktail in the 1960s. But he also had an impact on one particular brand of whiskey. When concertizing across the United States, Sinatra often sipped on some brown liquid from a rocks glass. That liquid was Jack Daniel's.

Sinatra had a friend named Angelo Lucchesi, a short, squat Memphis native who was hired in the 1950s by Brown-Forman as the brand's first marketing agent. At the time, Jack Daniel's was not a widely known brand. Lucchesi made sure Sinatra had a steady supply of Jack. "You keep making the booze," Sinatra wrote Lucchesi in 1976, "and I'll keep drinking it." The singer not only drank it onstage, he brought it with him to restaurants in his own traveling version of BYOB. In the early 1970s, a story went, Sinatra was in Athens and was dismayed when he ordered the whiskey at a certain bar only to be told they didn't carry it. To correct this baleful situation, Sinatra had a case sent to the bar with instructions that it be served gratis to any American who ordered it.

Sinatra supposedly drank his favorite whiskey the 3-2-1 style: a two-finger pour of whiskey with three ice cubes and a splash of water in a rocks glass. When he died in 1998, he was buried with a bottle of Jack. In 2016, the distillery honored Sinatra by releasing a special bottling called Jack Daniel's Sinatra Select.

FRANK'S WAY

This isn't really a cocktail. But who's going to argue with Sinatra?

2 ounces Jack Daniel's Tennessee Whiskey

Water

Add the whiskey to a rocks glass filled with three large ice cubes. Top with a splash of water.

1953

MAKER'S MARK

This is the most important bourbon ever created. Possibly. Very possibly. I'm not saying it's the best. I'm not saying it's the best made. I'm not saying it's the most popular, though it is very popular. I'm not saying anything other than that it's the most important. Important, meaning *impactful*. What was its impact? Enormous. Its impact was enormous.

The story belongs to the Samuels family, one of bourbon's great clans, who bought a distillery in Loretto, Kentucky, in 1953. At a time when bourbon was regarded as an affordable, humble spirit, Bill Samuels, Sr.—who came from a long line of distillers—decided it was time for whiskey to step up. He wanted to create a "premium" bourbon, which mainly meant more expensive, though he did want it to taste good, too. Maker's Mark hit the stores in 1958. Samuels didn't back down. The brand's tagline for many years was, "It tastes expensive . . . and is." It took a while, decades really, but eventually business took off.

Maker's Mark is a wheated bourbon, meaning its secondary grain is mellow wheat, not spicy rye. Samuels wanted a more approachable flavor profile that would not scare off novice consumers. As such, it may have led to the American tendency to equate wheated bourbon with high value, a belief cemented when Pappy Van Winkle, another wheater, became the holy grail for many drinkers in the twenty-first century. (Historically and aesthetically, the two brands belong together. Julian Van Winkle was an early mentor of Bill, Sr.)

It's debatable whether Maker's Mark would be as consistently popular as it is without its packaging, which is arguably the most distinctive in the business. The bottle, with its boxy base and long neck, is interesting enough, as is the hand-cut label with its unorthodox font usage. But it's the red wax that serves as its seal, a touch both old-fashioned and flamboyant, that makes it. Opening up a bottle of Maker's, pulling that string through the thick wax instead of just twisting a plastic cap, is as dramatic and romantic a feeling as breaking the emblem on a wax-sealed, handwritten letter.

These details were all the work of Margie Samuels, Bill's wife. Margie also named it Maker's Mark instead of something more obvious like Samuels Straight Bourbon. That name brought in the idea of craft, something that wasn't yet an aspect of modern bourbon distilling. All this makes Margie Samuels one of the best marketing minds in bourbon history. Margie also did a lot to make the Maker's Mark distillery an attractive property, a place someone might like to actually visit, thus anticipating a future where bourbon tourism was a thing. It is, indeed, one of the more beautiful distilleries in Kentucky, with its black buildings and red shutters, as close to central casting for a bourbon background as you're going to find. When you visit the distillery today, you have the opportunity to dip your own personal bottle in the hot custom red wax.

Maker's Mark passed through many hands beginning in 1981, when Hiram Walker & Sons bought it. It eventually landed with Jim Beam, which became Beam Suntory in 2014 and is Suntory Global Spirits today. The Samuels family, however, remains involved.

For years, nay decades, Maker's Mark made hay of the fact that it made only one thing. Why mess with perfection, they said. Why divert stock from a product the demand for which they couldn't even begin to satisfy. That was the party line. But in 2009, they came out with Maker's 46. It is made of the same stuff and made the same way, except for one critical aging fillip. When the bourbon is ready, the barrels are dumped of all liquid. Then, inserted into the barrels are seared (not charred) staves of French limousine oak. The booze is then poured back in and aged for two to four more months before being bottled at 46 percent ABV. The name supposedly derives from the fact that the approved formula took chief distiller Kevin Smith forty-six tries to get right. Like WD-40. It had a completely different bottle shape at first, to distinguish it. But the company later went back to the boxy bottle, because, why fix something that ain't broke?

BOURBON-MINT ICED TEA

AUDREY SAUNDERS, TONIC, NEW YORK CITY (2001)

"The Bourbon-Mint Iced Tea was my very first low-ABV creation," Saunders told me. The drink was on the lunch menu at Tonic. "I only utilized ¾ ounce bourbon in there so folks wouldn't get too hammered at lunchtime. The ¾ ounce bourbon and ¼ ounce Branca Menta provide just enough firepower to make the overall flavor profile pop, but not become overwhelmingly alcoholic."

¾ ounce Maker's Mark bourbon

¼ ounce Branca Menta

½ ounce fresh lemon juice

¾ ounce Simple Syrup (see page 23)

5 ounces black iced tea

Lemon wheel and mint sprig, for garnish

Combine all ingredients except the tea and garnishes in a mixing glass half filled with ice and quickly stir. Strain into a highball glass filled with ice. Top with the iced tea. Garnish with the lemon wheel and mint sprig.

GOOD NIGHT, IRENE

AUDREY SAUNDERS,
BEMELMANS BAR, NEW YORK CITY (2004)

This drink is a close cousin to the one above. "Good Night, Irene is a whiskey stinger, plain and simple," Saunders told me. "It is meant to be a nightcap, so I went with the easy-going Maker's, which utilizes soft winter wheat."

2 ounces Maker's Mark bourbon

1 ounce Branca Menta

Mint sprig, for garnish

Combine the liquid ingredients in a cocktail shaker and shake until chilled. Strain into a rocks glass filled with crushed ice. Garnish with the mint sprig.

1957
EVAN WILLIAMS

Bourbon and, to a lesser extent, rye is a name game. If you want to sell some bourbon, you slap an old dead guy's name on the label, preferably one that sounds old-timey and smacks of Anglo-Saxon heritage and, if possible, has something to do with the history of your company. If you can't swing that, a name having something to do with bourbon history in general will do. And, in the best of all worlds, put the word "old" in front of the name. Jack Daniels, Jim Beam, Elijah Craig, and many more follow this playbook to one degree or another.

The second bestselling bourbon in the world is Evan Williams, which is produced by Heaven Hill. The bourbon was introduced in 1957, but the labeling made it seem like the juice had been on shelves for a century. Ads claimed it was made "from a formula originated in 1783" and that the whiskey was made at a place called the "Old Evan Williams Distillery" in Bardstown.

Williams was a Welsh immigrant who settled in Louisville and began distilling in 1783. Heaven Hill asserts on its label that Williams was America's first distiller of bourbon. That boast was on the bottles as far back as 1958 (along with the tagline "So M-e-l-l-o-w. So S-m-o-o-t-h").

Evan Williams is regarded by most bourbon experts and fans as one of the best value brands on the shelf. The flagship brand—whose familiar boxy bottle and black label made its debut in 1965—used to be advertised as a seven-year-old bourbon but no longer is. It's believed to be around four years old now. Still, at under $20 a bottle, no one is complaining. The Evan Williams Single Barrel, which is much better, is also thought to be a steal at $40.

FINELY
CRAFTED
world-renowned by
Bourbon lovers.
EXTRA AGED IN OAK
Since 1783
Evan
Williams
KENTUCKY'S 1ST DISTILLER
KENTUCKY
Straight
BOURBON
Whiskey
43% ALC/VOL

REXTON RAWLSTON

CHARLES CERANKOSKY, ROCHESTER, NEW YORK (2016)

Cerankosky is the leading cocktail bar owner in Rochester, New York, and the founder of the Rochester Cocktail Revival, an annual booze convention.

- 1½ ounces Evan Williams 1783 Bourbon
- ¾ ounce crème de cassis
- ¼ ounce Michelberger Mountain schnapps
- ¼ ounce fresh lemon juice
- 3 dashes of Abbott's bitters
- Cava
- Cocktail cherry, for garnish

Combine all ingredients except the cava and cherry in a cocktail shaker half filled with ice and shake until chilled. Strain into a Nick & Nora glass. Top with the cava, then garnish with the cherry, skewered on a pick.

1964

THE BOURBON ACT

Got rules? Bourbon does. Lots of them. Bourbon is, in fact, one of the most regulated spirits, thanks to the 1964 resolution passed by Congress, which simultaneously recognized bourbon as a "distinctive product of the United States" and decreed the stuff could only be made in the US of A (including the District of Columbia and Puerto Rico). It also rolled out this laundry list of requirements:

Bourbon must be made from a grain mixture that is at least 51 percent corn.

Bourbon must be aged in new, charred oak containers, those containers being barrels.

Bourbon must be distilled to no more than 160 proof—that is, 80 percent alcohol.

Bourbon must enter the barrel for aging at no more than 125 proof (62.5 percent alcohol).

Bourbon must be bottled at 80 proof or more (40 percent alcohol by volume).

The government gave bourbon a little slack on aging requirements. There is no minimum age specified. Distillers can release liquid that's only been aged a few months. If they want it to be called "straight" bourbon, however, it must be aged a minimum of two years.

Though all the regulations might seem onerous, this was good advertising for bourbon. It was now America's official spirit. It was like those products in Great Britain that get a stamp on the label saying the House of Windsor likes and uses the stuff. Not many people paid attention at the time, including the press; perhaps they correctly assumed that the legislation was little more than a publicity stunt. And it didn't seem to have swayed the public much either; in the years to come, American whiskey did nothing but lose market share to clear spirits. But it proved a boon in the long run, lending the industry a certain prestige when it came roaring back in the twenty-first century.

1972
THE DARK AGES

The "Dark Ages" of American drinking is a coinage of the cocktail revival. It refers roughly to the 1970s through the 1990s and denotes an era when people drank badly—bad spirits, bad cocktails, badly made. Make no mistake, it's a highly judgmental phrase. The people who invented it thought folks back then were drinking wrong: vodka instead of gin; mixto tequila instead of 100 percent agave; syrupy, chemically laced liqueurs instead of the heritage amari and bitters of Europe; Kamikazes and Disco drinks instead of Manhattans.

A bourbon distiller of the era might have called it something different, perhaps the Age of Survival. Clear spirits were eating their lunch. Everyone was drinking vodka and tequila. Bourbon was an old drink meant for old folks. Hell, most of the brands had the word "old" in the title. Age and history were bourbon's stock-in-trade, the things that made it respected and valued for decades. That didn't play anymore. America's youth found their country's history suspect and unsavory. Just as the young hippies of the late 1960s decided they'd rather do other drugs than sip mom and dad's Martinis, the young drinkers of the late 1970s decided bourbon was for country clubs and gentlemen's libraries.

Rye was too weak to fight back at this point. As for bourbon, its general marketing position was "if you can't beat 'em, join 'em." Many brands lowered their alcohol content to mimic the so-called smoothness of vodka. Others tried to actually look like vodka, stripping the whiskey of its color. One of these was Frost 8/80, a curious creation of Brown-Forman. It was an eight-year-old bourbon in which all color had been filtered out and the proof brought down to eighty (8/80, get it?).

The genre was called "light whiskey," with federal laws defining that category going into effect in 1972. The spirits in question were aged in mature oak barrels, leading to less-woody whiskey. Four Roses had a brand called Four Roses Premium, which was advertised with the tagline, "Prepare to be underwhelmed." Old Crow came out with Old Crow Light, also 80 proof. Crow's line was, "You've tried some whiskeys that growl. Now try a new kind of whiskey. Crow Light. It Whispers." The label shows two highballs filled to the brim with ice. No neat drinking here. Other new brands were called White Balloon, Northern Light, Skylark, and Honey-Go-Lite. Yup. It all sounded awful.

None of these salvos hit their mark. Drinkers saw the ad campaigns for what they were: desperate and corrupt. Most brands lasted only a few years, though the general ABV of most bourbons stayed low.

BOURBON SLOE GIN FIZZ

The Sloe Gin Fizz was arguably one of the classier choices of the decidedly unclassy cocktail era of the late twentieth century. Here's a bourbon version of the drink using Fords, a recent, upgraded rendition of the liqueur that defines the cocktail.

2 ounces bourbon

1 ounce Fords Sloe Gin

¾ ounce fresh lemon juice

2 ounces seltzer water

Lemon slice and cocktail cherry, for garnish

Combine the bourbon, sloe gin, and lemon juice in a cocktail shaker half filled with ice. Shake until chilled. Strain into a Collins glass filled with ice, add the seltzer, and garnish with the lemon slice and a cherry.

1976

FLAVORED BOURBONS

Once upon a time, in the late twentieth and early twenty-first centuries, if you mentioned the flavors honey, cherry, cinnamon, apple, or maple syrup in connection to spirits, one word would come to mind: vodka. Flavored vodkas were legion. But by the 2010s, bourbon was flirting with flavors that went beyond the caramel, vanilla, and oak imparted by the barrels. Distillers began sugaring and flavoring whiskey. Wild Turkey Liqueur, a honey-infused bourbon, was an early entry, introduced in 1976 and later reformulated as American Honey in 2006. But with Red Stag, the successful cherry-flavored bourbon released by Jim Beam in 2009, the dam burst. After that, there were flavored American whiskeys from the bourbon distillers Heaven Hill (Evan Williams Honey Reserve and Cherry Reserve) and Brown-Forman (Early Times Fire Eater, which tastes of red-hot cinnamon, and apple-flavored Blind Archer), and the Tennessee whiskey giant Jack Daniel's (Tennessee Honey).

The advent, and popularity, of flavored bourbon over the last decade has been a sore spot for purists. But if bourbon was indeed booming—and it was—these bottles are part of the reason. In 2023, flavored whiskeys accounted for 24 percent of total whiskey volume.

Whether the flood of new flavors was good or bad for whiskey's image depended on whom you talked to. People in the industry say flavored whiskey is a gateway drink that will introduce novice drinkers to the spirit. Eventually, the logic goes, their tastes will evolve, and they will make the leap to straight whiskey. "The starting point used to be Bourbon and Coke, or Bourbon and Ginger," Dave Pickerell, the former master distiller at Maker's Mark, once told me. "This is just one extra step further down the trail."

But to some purists, the products have more in common, commercially and aesthetically, with popcorn vodka than corn whiskey.

Though industry officials don't like to talk about it, it's no secret that one initial aim of the bottlings was to bring women into the whiskey fold.

Though the flavored-whiskey shelf at the liquor store is getting crowded, Chris Morris, the master distiller at Brown-Forman, once said to me that he doesn't think whiskey is headed down vodka's road to perdition. Vodka is a blank slate; whiskey has a character that must be harmonized with. "I think that will give us a more rational, tighter range of expressions," he said.

ES OLD-FASHIONED

JACKSON CANNON,
EASTERN STANDARD, BOSTON (2025)

Cannon is a longtime leader in the Boston bartending world and is mostly associated with the popular restaurant and bar Eastern Standard. For the house Old-Fashioned, he splits the spirit base between bourbon and flavored bourbon. For the latter, he uses a product from Mad River Distillers in nearby Vermont.

- 1 ounce Elijah Craig Cask Strength Bourbon
- 1 ounce Mad River Double Tapped Maple Bourbon Whiskey
- ½ ounce Bénédictine
- Dash of Angostura bitters
- Orange twist, for garnish

Combine the liquid ingredients in a mixing glass half filled with ice. Stir until chilled and strain into an Old-Fashioned glass filled with ice. Garnish with the orange twist.

1984

SINGLE-BARREL AND SMALL-BATCH

"Single barrel" is a common bourbon category today, a term every bourbon drinker knows, even if they don't always understand it. Blanton's was the first, debuting in 1984. It was the creation of Elmer T. Lee, a protégé of Albert Bacon Blanton and an employee of what would become Buffalo Trace distillery. He saw a future for premium bourbon when many did not.

Lee plucked select barrels for their excellence and put the contents in a fine, fancy package topped by an unmistakable metal stopper shaped like a horse and jockey. He then slapped an unheard-of-at-the-time $25 price tag on it. Japan was the target market, but it slowly caught on in the United States. Today, you're lucky if you can find a bottle.

"Elmer T. Lee realized this type of whiskey could save the struggling category and asked the other distillers to follow suit," whiskey expert Noah Rothbaum once told me.

Single-barrel expressions started to proliferate in the aughts. Now, nearly every big-name bourbon distillery has one.

The first thing to know about the term *single barrel* is that it isn't regulated. This leads directly to the second thing you need to know, and that's that single-barrel bourbon is a misnomer in most cases, not to say an outright fib. That's because it doesn't come from a single barrel. That would be silly and economically impossible.

This isn't to say single-barrel bourbons aren't good. Most of them are noticeably better than the distillery's standard offerings. This is because producers are more careful in their selection of barrels, choosing only what they consider the best. (All barreled whiskey ages differently, depending on many factors, including the barrel being used, its place in the rickhouse, the weather it has been exposed to over the years, and other mysterious things.)

There's another term that is regularly tossed around with single barrel, and that's *small batch.* This is also unregulated and thus open to abuse. And it's even more vague. Generally speaking, a small-batch bourbon is drawn from a blend of a couple hundred or so barrels. So the distiller is being selective, but not as selective as with single barrel. If the term ever meant much, it is generally regarded as essentially meaningless today. But, again, that doesn't mean the bourbons sold under that name are bad. Usually, they're better. Distillers still have to match their hyperbole with quality, even if the language is essentially nonsense. Otherwise, the customer won't come back.

MIDNIGHT TRAIN

LUCINDA STERLING, MIDDLE BRANCH, NEW YORK CITY (2014)

Sterling, a protégé of Sasha Petraske, who kicked off the global cocktail renaissance when he opened his New York bar Milk & Honey, invented this Old-Fashioned riff when she ran the respected Manhattan cocktail bar Middle Branch in the 2010s. The original recipe doesn't exactly specify small-batch bourbon, but its use certainly couldn't hurt.

2 ounces small-batch bourbon

½ ounce Giffard peach liqueur

2 dashes of Angostura bitters

Lemon twist, for garnish

Combine the liquid ingredients in a mixing glass half filled with ice and stir until chilled. Strain into an Old-Fashioned glass filled with one large ice cube. Garnish with the lemon twist.

1987

BULLEIT

The Story of Bulleit is the story of modern bourbon in a nutshell. The company was founded in 1987 by a lawyer, Tom Bulleit, with a surname that was possibly related to the nineteenth-century origins of America's greatest spirit. (Tom claimed his great-great-grandfather Augustus Bulleit made bourbon from 1830 to 1860.) He launched a modern version of his family whiskey, perhaps based on an old recipe by Augustus, perhaps based on a long history, and perhaps not. It was then presented to a media and consumer base thirsty for such nostalgic stories, and to a bartending community thirsty for quality liquid (that it was a high-rye bourbon helped its fortunes, as bartenders liked that style). Bulleit pounded the pavement making his whiskey's case and eventually got bought by a corporation (first Seagram, then Diageo) looking to pick up some authenticity points in the bourbon world. It then quickly became an international brand.

Despite its unprovable and improbable backstory; despite family infighting (both father and daughter Bulleit were eventually sidelined); despite a transparently manufactured persona (the bottle reads "Frontier whiskey"); despite the fact that for decades Bulleit didn't actually make its own whiskey but sourced it from Four Roses and others—despite all that, it worked. Bulleit became big. You see it on most back bars. It didn't hurt that the liquid was good.

Bartenders did their part, too, as they always have during the past quarter century. The brand took off in San Francisco, where Stephen Beal, a whiskey business veteran who worked with Diageo, and Bulleit himself hand-sold many bottles, marketing it as an affordable craft spirit. "Bulleit began as a tiny brand and our strategy was engagement on a one-on-one basis with our 'partners in chemistry,'" Beal once told me. "Jon Santer was one of the first."

Bartender Santer was working at Bruno's, the jazz club in San Francisco's Mission District, where his boss did the liquor buying. "One week I found myself with a case of Bulleit bourbon," Santer told me, "a whiskey no one was going to call for, because no one knew what it was. I started experimenting with it because I had it and needed to sell it."

Santer used the Bulleit to make a new Manhattan riff called a Revolver, made of bourbon, coffee liqueur, orange bitters, and a flamed orange peel. But it wasn't until the opening of Bourbon & Branch in the city's Tenderloin neighborhood in 2006 that the Revolver gained a real foothold. Santer was on the opening bar staff, and his Revolver was the only original cocktail on the debut menu not created by bar director Todd Smith.

Bulleit has its own distillery now and distills its own product.

REVOLVER

JON SANTER, BRUNO'S, SAN FRANCISCO (2004)

This clever twist on a Manhattan was a hit in the early years of the cocktail renaissance in San Francisco.

- 2 ounces bourbon
- ½ coffee liqueur
- 2 dashes of orange bitters
- Flamed orange twist, for garnish

Combine the liquid ingredients in a mixing glass half filled with ice and still until chilled. Strain into a chilled coupe. Garnish with a flamed orange twist.

THE NINTH WARD

ROBERT "BROTHER CLEVE" TOOMEY (2008)

Brother Cleve, who died in 2022, was the godfather of the Boston craft cocktail scene. He considered this drink to be his best-known invention. He created it for Tales of the Cocktail convention in 2008. About the Ninth Ward, he once told me, "I wanted to create a drink for the event that would have some sort of New Orleans and Boston connection. So my idea was to take the Ward Eight, the best-known drink created in Boston, and turn it into a tropical cocktail for New Orleans." The Ninth Ward in New Orleans is home to Brother Cleve's favorite dive in that city, Saturn Bar. He went so often, the bar took messages for him.

- 1½ ounces bourbon, preferably Bulleit
- ¾ ounce fresh lime juice
- ¾ ounce falernum
- ½ ounce St-Germain elderflower liqueur
- 2 dashes of Peychaud's bitters
- Orange slice, for garnish

Combine all liquid ingredients in a cocktail shaker half filled with ice and shake until chilled, about 15 seconds. Strain into a chilled cocktail glass. Garnish with the orange slice.

1990s

NON-DISTILLING PRODUCER

No term in the bourbon and rye world has undergone a more complete transformation than Non-Distilling Producer (NDP for short). Non-Distilling Producer is a confusing, wordy phrase for people who source, rather than make, their whiskey; i.e., Bob sells Bob's bourbon, but Bob doesn't make bourbon; he buys it from Bill, who does, and labels it Bob's.

The public didn't have to deal with this term much in the past, even though sourcing whiskey is a very old practice. That's because people didn't used to feel terribly compelled to talk about where their bourbon and rye came from or who made it. But beginning in the 1990s, when bourbon and rye began to heat up as commodities, NDPs became more of an issue as more and more people wanted to get in on the action. Striking out as a whiskey distiller is an expensive and lengthy proposition because you not only have to build a distillery, you also have to wait years before the whiskey you distill is ready to bottle and sell. A speedy and economical stopgap is to buy surplus whiskey from the guys who are already making it, then bottle it, brand it, and sell it as your own. Bulleit, Jefferson's, and Michter's were among the first to do this in the modern age.

In the early years of this boomlet, in the nineties and aughts, NDP was a dirty word and its practitioners were thought to be a little shifty. Only real distillers with real distilleries made real whiskey, the thinking went. Everyone else was a carpetbagging parvenu.

Part of this was faulty perception on the part of the press and public. But part of this unsavory rep was the fault of some of the NDPs, who were less than transparent about where their whiskey came from or whether they made it at all. Templeton Rye was a notable example. To hear the producers tell it, this whiskey was Al Capone's favorite back in the day and was made in tiny Templeton, Iowa, where it was relaunched in 2006. But no rye was made in Iowa. The juice was bought at the huge MGP distillery in Lawrenceburg, Indiana.

Whistlepig pulled a similar stunt, acting like its rye came from its base in Vermont, when it was really shipped down from Canada. By the 2020s, NDPs were no longer regarded as shysters. They were principled artists who were following a vision and sold some of the best whiskey on the market. Also, some of the NDPs started making their own whiskey, fully transitioning from buyers to distillers. Michter's and Bulleit are among these. So is Templeton, which rolled out its first homemade whiskey in 2024. It's even made in Templeton, Iowa.

THE LEATHER RECLINER

MIKE VACHERESSE, TRAVEL BAR, BROOKLYN (2020)

Travel Bar is arguably the finest whiskey bar in New York City. But it also serves great cocktails. The Leather Recliner started life as a barrel-aged cocktail Vacheresse created during the Covid-19 pandemic. "The cocktail is big and luscious," he told me, likening the flavor to sitting in a leather recliner in a dark-paneled room.

- 1¾ ounces high-proof rye (preferably 115 proof New Riff Single-Barrel Rye)
- ½ ounce Luxardo Amaro Abano
- ¼ ounce Suze
- Cocktail cherry, for garnish

Combine all liquid ingredients in a mixing glass half filled with ice and stir until chilled. Strain into a double Old-Fashioned glass over one large ice cube. Garnish with the cherry.

1996
WOODFORD RESERVE

Woodford Reserve was one of the first modern bourbons to put on evening clothes. Released in 1996, its vessel was not round. It was square and flat, like a big bottle of perfume. The high-end profile was intentional. Brown-Forman, which produces Woodford, wanted a brand that would be held in the same esteem as the best Scotch in the world.

Brown-Forman bought an old distillery south of Frankfort—land that was formerly the home of distilling pioneer Oscar Pepper, his son James Pepper, and Dr. James Crow, the chemist they hired to distill their whiskey (see page 57). That's a pretty good pedigree, and it lent Woodford a patina of historical authenticity. And the old buildings, once spruced up, were dramatically atmospheric. That was important, because Woodford was one of the first bourbon brands to welcome visitors, thus becoming a pioneer of bourbon tourism. In 2024, Woodford restored the old Pepper house, which had been vacant for decades, the ruins sitting on a hill overlooking the distillery. It is now home to a fancy tasting room where private barrel bottlings are selected by the fortunate few.

Woodford established itself as a premium bourbon brand fairly early and has been regarded as such ever since, even though the premium category has become significantly more crowded. Today, a deep-pocketed customer at a whiskey bar would probably regard Woodford as a relatively affordable gateway tipple into the world of classy bourbon.

Woodford also pioneered the idea of limited special releases under its Master's Collection umbrella: whiskeys finished in unusual barrels that used to hold Chardonnay and Pinot Noir and such. All these innovations came under the guidance of Chris Morris, the long-serving master distiller who succeeded Lincoln Henderson in the job in 2003. Morris was the face of the brand until he was succeeded by Elizabeth McCall in 2023. (The library in the old Pepper house was named after Morris, a fitting tribute since the soft-spoken, bespectacled Morris often came off more like a librarian than a distiller. As odd as it is to say, very few bourbon distillers in Kentucky wear glasses, at least in public.)

Woodford has been good about integrating itself into the bourbon-world fabric. It has been the official bourbon of the Kentucky Derby since 1999, when it was only three years on the market. It is poured into countless Juleps, including some in special julep cups that fetch up to $5,000, part of an annual charity program.

GOING UPTOWN

**STEVEN FERREIRA,
BAR BELLY, NEW YORK CITY (2024)**

This Manhattan-style cocktail expresses Ferreira's experience of growing up Dominican in New York. Mamajuana, a popular ad hoc alcoholic beverage that has been drunk in the Dominican Republic for many years, is typically a mixture of rum, wine, honey, and spices. Ferreira makes his own, which he uses as a modifier in this Manhattan riff. "Going Uptown is all about honoring the traditional values of mamajuana—family, tradition, and its medicinal roots—by reimagining it in today's cocktail culture," Ferreira told me. Mi Pana Mamajuana grew out of that original recipe. He provides a simplified version of his own mamajuana here.

- 2 ounces rye, preferably Woodford
- 1 ounce Mi Pana Mamajuana (or use homemade; recipe follows)
- 2 dashes of Angostura bitters
- 4 dashes of Fee Brothers Peach Bitters
- Flamed orange twist, for garnish

Combine the liquid ingredients in a mixing glass half filled with ice. Stir until chilled. Strain into a rocks glass filled with one large ice cube. Express the flamed orange twist over the drink and discard.

ALTERNATIVE MAMAJUANA

8½ ounces Dominican white rum	In a large bowl, stir together the ingredients to combine. Store in a glass bottle.
10 ounces red wine	
2½ ounces Amaro Averna	
2 ounces Strong Honey Syrup (recipe follows)	
1 ounce Angostura bitters	

STRONG HONEY SYRUP MAKES ABOUT 1 CUP

¾ cup clover honey	In a medium bowl, stir together the honey and water. Let cool and store in a glass bottle.
¼ cup hot water	

COMMONWEALTH COCKTAIL

CHARLES JOLY,
CHURCHILL DOWNS, KENTUCKY (2017)

Joly created this cocktail to be served at Churchill Downs for the 2017 Kentucky Derby. "I had the opportunity to create signature cocktails for the event for several years," Joly told me. "I attended in person and headed up a specialty bar during the weekly events. We popped up in the Turf Club, Clubhouse, Millionaire's Row, and even the uber-exclusive Speakeasy."

- 1½ ounces Woodford Reserve bourbon
- ½ ounce Giffard grapefruit liqueur
- ¾ ounce fresh lemon juice
- ¼ ounce Honey Syrup (see page 69)
- Dash of Bittered Sling Grapefruit & Hops bitters (orange bitters will do in a pinch)
- 1½ ounces chilled dry sparkling wine
- Long grapefruit twist, for garnish

Combine all ingredients except the wine and grapefruit in a cocktail shaker half filled with ice. *Gently* shake to quickly combine and chill. (With any cocktail that is getting soda or another lengthener that will contribute dilution, you don't want to double-dilute and wash out the cocktail.) Strain into a chilled flute and top with the sparkling wine. Garnish with the grapefruit twist.

1996

PAPPY VAN WINKLE

Even if you're a teetotaler who has lived in a cave for the past twenty years, this is probably still the one bourbon that would ring a bell. Allocated, hard to find, and a wallet-emptier when you do find it, the Van Winkle line fostered a hunting mentality in bourbon drinkers and helped elevate bourbon into a luxury commodity on par with rare Bordeaux.

In an era of great bourbon, a significant contingent of fans have decided that this brand is the best, the ne plus ultra, the nobody-does-it-better prestige juice of all time. It isn't. It's really good, if you can find it, and if you can afford it. But a lot of other bourbon bottlings are just as good and easier to locate, including some of those made by Buffalo Trace, the distillery that produces Pappy Van Winkle.

The Van Winkles are real people, a family that share a surname with Vanilla Ice. And their whiskey roots go far back. But they don't own a distillery and haven't for decades. Julian Van Winkle III merely shepherds a brand.

Scarcity is what drives up the value of Pappy (as it's known by all). There's only so much of it. Several thousand bottles are released once a year. And what there is, is "allocated" to only the most loyal and powerful retail and restaurant accounts. That means the likelihood of your finding a bottle of 15-, 20-, or 23-year Pappy Van Winkle's Family Reserve at your local liquor store is next to nil.

Given the sui generis marketing strategy the brand has pursued for the last thirty years, it only makes sense that Julian "Pappy" Van Winkle, Sr., the old guy chomping a cigar on the label, was himself a salesman. He started working for W. L. Weller & Sons in 1893 when he was still a teenager. In 1910, he and Weller bought the A. Ph. Stitzel Distillery in Louisville. Just before Prohibition hit, they started making a brand called Old Rip Van Winkle. The Van Winkle family held on to that name even as Old Rip wasn't made any more, and even after they sold the distillery in 1972. That was remarkably prescient. When Julian Van Winkle, Jr., reintroduced the brand in 1996, he used old stock bought from the new owners of his old distillery. The 20-year-old cost a hefty $70. It quickly won a reputation, winning awards at the World Spirits Competition, scoring a ninety-nine on a scale of one hundred. Over time, chefs like Anthony Bourdain, Daniel Boulud, and David Chang proclaimed Pappy the best bourbon. (For whatever reason, whiskey buyers are more prone to take the advice of chefs than of bartenders or distillers when choosing spirits.)

The sourcing arrangement with Stitzel obviously couldn't go on indefinitely, so, beginning in 2002, Pappy started being distilled at Buffalo Trace, which is owned by the

New Orleans–based Sazerac Company. Once the Stitzel juice ran out and the Buffalo Trace stuff took its place, people were just buying a name and a recipe. It was no longer rare because it was *actually* rare; it was rare on purpose, because Buffalo Trace only made so much.

These days, a bottle will go for thousands of dollars on the secondary market; a one-ounce dram in a bar sells for hundreds. The arguments about whether it's that good or worth it are all academic at this point, because Pappy is now permanently a bourbon for the 1 percent.

(There is no cocktail recipe for this chapter. If you're fortunate enough to lay your hands on a bottle of Pappy, I suggest you savor it straight or on the rocks.)

1997

MICHTER'S

This old Pennsylvania distilling name got a new lease on life in the 1990s under new owners who sourced, rather than made, their whiskey. Joseph Magliocco, the owner of Chatham Imports, a Harvard- and Yale-educated New Yorker, is about as unlikely a figure as you're going to find in the Kentucky whiskey world. He noticed that the trademark for Michter's, a defunct Philadelphia brand of rye, was up for grabs. He used to sell Michter's when he had a sales job at his family's wine and spirits business. So, perhaps feeling a little bit sentimental, or perhaps spying an opportunity, he forked over $245 and bought it. He then set about reviving the brand. (There was no Old Man Michter in the whiskey's past. The brand was created in the mid-twentieth century by Lou Forman, an ad exec. The name was a portmanteau of his sons' names, Michael and Peter.)

Magliocco's tenure as the steward of Michter's began as a non-distilling producer (see page 142). He sourced whiskey from others until he was able to release his own aged liquid. He also left Pennsylvania behind and moved the Michter's name to Kentucky, where, by that point, all the other major rye whiskeys in the country were being made. He also decided that Michter's would make a bourbon for the first time in the label's history. Plus, he wanted those bourbons and ryes to be a minimum of ten years old. The first distiller of this reborn Michter's was Willie Pratt, an industry veteran who spent decades at Brown-Forman. It was just another instance of Magliocco spying an undervalued asset in the American whiskey scene (like rye, like aged spirits, like the name Michter's).

They are now producing whiskey at their own distillery in Shively, Kentucky, which opened its doors in 2012; and, as of 2018, they are also distilling on a smaller scale at their Fort Nelson facility on Main Street in Louisville. (Michter's stopped sourcing liquid in 2015.) Together, the two buildings represented a huge investment and show of faith in the Kentucky bourbon and rye industries. That investment paid off. Michter's releases are among the most prized bottlings in American whiskey, with a flavor profile that has remained consistent whether it was sourcing whiskey or contracting it from other distillers. In addition to the 10-year-old expressions, there are also occasional 20- and 25-year-old whiskeys that are well worth seeking out.

Michter's is an innovator in other unseen ways as well. Magliocco may have had something to do with the fact that almost all modern rye whiskey sports a green label. In the late 1990s and early 2000s, when the Michter's team was working on the design

for what would become its rye, Magliocco gravitated toward green. "Green went with the feel we wanted," he once explained to me. "When I think of red, I think of sweeter things. Green is a little spicy, a little herbaceous. Rye is also a winter crop and green can be a sort of winter color."

FORT NELSON CRUSTA

DAVID WONDRICH,
THE BAR AT FORT NELSON,
LOUISVILLE, KENTUCKY (2019)

When Joe Magliocco, the owner of Michter's, opened the Fort Nelson Distillery in downtown Louisville, he asked cocktail historian David Wondrich to design the cocktail list for the in-house bar. Wondrich came up with this old-school twist on the Crusta, a mid-nineteenth-century sour. The drink has become the bar's signature cocktail.

1¾ ounces Michter's US*1 bourbon

¼ ounce yellow Chartreuse

¾ ounce fresh lemon juice

¼ ounce Demerara Syrup (see page 43)

¼ ounce honey

3 dashes of creole bitters

Long lemon twist (a.k.a. horse's neck), for garnish

Lemon wedge, for the glass

Granulated sugar, for the glass

Wet the rim of a cocktail glass with the lemon wedge, then coat the rim with granulated sugar. Combine the liquid ingredients in a cocktail shaker half filled with ice and shake until chilled. Strain into the cocktail glass. Garnish with the long lemon twist, arranging the "horse's neck" in a series of stylish hoops that extend within and without the liquid.

1999

BOURBON TRAIL

Bourbon Tourism. That's a phrase you never heard in the nineteenth century, when the bourbon industry was getting on its feet. Not in the twentieth century either, when bourbon was a household word, even as its fortunes constantly waxed and waned.

But at the tail end of the latter century, a seed was planted that grew into the mighty capitalistic oak that is today's bourbon tourism industry. It is called the Kentucky Bourbon Trail, the brainchild of the Kentucky Distillers' Association, which was founded way back in 1880. The first steps down that path were made in 1999. At the beginning, there were all of seven distilleries on this trail, which was a pretty good percentage given that there were only nine bourbon distilleries at the time. In many cases, this marked the first time civilians could step inside these places of business. There was a passport that was stamped with each distillery visit. Growth was slow. In 2013, there were eight stops. But today there are sixty spanning Louisville, Lexington, Bardstown, and beyond.

On the trail, you can meet a distiller, tour a distillery, tour a rickhouse, thieve some juice from a bourbon barrel, stay at an affiliated hotel, eat at an in-house distillery cocktail bar or restaurant (they all have one now), taste through a flight of bourbons, take a cocktail class, bottle your own personal bourbon, and, of course, exit through the gift shop.

In 2024, the Bourbon Trail's twenty-fifth year, the trail received a record 2.7 million visitors from all fifty states and more than twenty countries, spending thousands of dollars per head.

At this point, rye whiskey is starting to get jealous of all those tourist dollars, but there are problems with creating a Rye Trail. Most rye is made by bourbon distillers, who already have a trail. And rye distilleries are not concentrated in one state, like Kentucky. But Beam Suntory is taking a run at a Pennsylvania Rye Trail with the Old Overholt museum in West Overton, Pennsylvania. Give them twenty-five years and they may come up with something.

BEHAVE YOURSELF

LINDSAY MATTESON,
HELL OR HIGH WATER, LOUISVILLE, KENTUCKY (2025)

This drink was invented at Hell or High Water, one of the best new cocktail bars in Louisville. Angel's Envy joined the Bourbon Trail in 2017, becoming the tenth member distillery on the trail.

1½ ounces Angel's Envy bourbon

¾ ounce Giffard banana liqueur

¾ ounce fresh lime juice

½ ounce Vanilla-Orange Cordial (recipe follows)

Dash of Bittermens Tiki Bitters

Mint sprig, for garnish

Freshly grated nutmeg, for garnish

Combine the liquid ingredients in a cocktail shaker half filled with ice. Shake until chilled. Strain into a double Old-Fashioned glass filled with ice. Garnish with the mint sprig and freshly grated nutmeg.

VANILLA-ORANGE CORDIAL

1 cup sugar

1 cup orange juice

1 vanilla bean, split

In a medium saucepan, combine the ingredients and simmer over low heat until the sugar has dissolved and the syrup smells like vanilla, about 15 minutes. Strain, discarding the solids, and let cool. Store in an airtight container in the refrigerator for up to 1 month.

2000
COCKTAIL RENAISSANCE

You can't tell the story of the bourbon and rye revival without telling the story of the parallel cocktail revival.

For a long time, in the late decades of the twentieth century, bourbon didn't get much help from cocktails. Old-timers ordered Manhattans and Old-Fashioneds. One day a year, bourbon was needed for Mint Juleps. Rye, meanwhile, didn't get a single assist from cocktails outside of New Orleans, where people still drank Sazeracs. Everything that went into cocktails, be it in fern bars or discos, was clear: vodka, flavored vodka, tequila, rum, and, occasionally, gin.

Then a few iconoclastic bartenders rediscovered classic cocktails, most notably Dale DeGroff, who was mixing it up uptown at the Rainbow Room in Manhattan, and Sasha Petraske, who was mixing it up downtown at a hole-in-the-wall called Milk & Honey. A lot of young bartenders followed their example. And guess what they suddenly needed to make authentic versions of the pre-Prohibition cocktails they wanted to serve? Bourbon and rye.

They were stirring up exacting versions of Manhattans and Old-Fashioneds, ones that young people were drinking. Blended and bottom-shelf whiskeys were out; the best bourbons and ryes were in. And there were other odd, old whiskey drinks coming out of the woodwork, drinks no one had heard of, or, if they had, had completely forgotten: Boulevardier, Brown Derby, Brooklyn, De La Louisiane, Deshler, Diamondback, Old Pal, Remember the Maine, Ward Eight, Whiskey Smash, Vieux Carre, and New York Sour. It was like someone had revived a lost language. Old Overholt and Old Grand-Dad were suddenly going into Old-Fashioneds.

What's more, these bartenders were inventing new cocktails that required bourbon and rye. (Who invented *new* cocktails?!) They had names like Gold Rush, Paper Plane, Kentucky Buck, Left Hand, Revolver, Greenpoint, Little Italy, Trinidad Sour, Red Hook, American Trilogy, Silver Lining, Benton's Old-Fashioned, Bourbon Renewal, Conference, and Black Manhattan.

Bourbon professionals love to sit in their rocking chairs and cluck about how they like their whiskey straight and that's the best way to drink it. But none of them can deny that cocktail bars and bartenders helped them buy that rocking chair—and probably the porch and the house attached to it.

GODDESS OF THE BRONX

DICK BRADSELL,
SHARED BY BEA BRADSELL (CIRCA 1989)

Dick Bradsell, who died in 2016, was the godfather of the London cocktail revival. He created several modern classic cocktails, including the Espresso Martini and the Bramble. This is a lesser-known invention of his. It was created for his friend, the artist Rita Keegan, who is from the Bronx. It was his own take on the classic Bronx cocktail—he hated the original (made of gin, dry and sweet vermouths, and orange juice) and believed the Bronx deserved a better drink. Bea Bradsell is Dick's daughter.

- ¾ ounce Maker's Mark bourbon
- ¾ ounce Myers's Rum
- ¾ ounce framboise liqueur
- ¾ ounce fresh lime juice
- Dash of Angostura bitters
- Barspoon of Simple Syrup (see page 23)
- Lemon slice, for garnish
- Orange slice, for garnish

Combine all ingredients except the garnishes in a cocktail shaker half filled with ice. Shake until chilled and strain into a double Old-Fashioned glass filled with ice. Garnish with the lemon slice and orange slice.

ANOTHER OLD PAL

DALE DEGROFF (2023)

DeGroff is the godfather of the American cocktail renaissance, having kicked it off while working behind the bar at The Rainbow Room from the late 1980s through the 1990s. More recently, he launched his own line of spirits, including the bitter aperitivo used here in this riff on the classic Old Pal cocktail.

- 1½ ounces rye, preferably Woody Creek
- 1 ounce DeGroff Bitter Aperitivo
- ½ ounce Dolin Dry Vermouth
- ½ ounce Martini Riserva Speciale Ambrato Vermouth
- Orange slice, for garnish

Combine the liquid ingredients in a mixing glass half filled with ice and stir until chilled. Strain into a double Old-Fashioned glass over one large ice cube. Garnish with the orange slice.

2000s

WOMEN IN WHISKEY

In 2013, bourbon writer Fred Minnick wrote a book called *Whiskey Women*. He told the stories of pivotal, but unsung, women in the bourbon business, such as Margorie Samuels, who was responsible for the name and image of Maker's Mark, and Peggy Noe Stevens, a master taster for Brown-Forman.

It was a necessary book at the time, because the bourbon industry was dominated—as it had been for two centuries—almost solely by white men. While that is still largely the case, things have changed for the better in the decade or so since that book was published.

While doing research for this book, I took a few trips to Kentucky. These were my first visits there since before the pandemic. At each distillery, I encountered women in positions of authority. There is Uncle Nearest in Kentucky, founded by Fawn Weaver, who hired Victoria Eady Butler as her master blender. Butler is also the great-great granddaughter of Nearest Green, the first known African American master distiller and the namesake of the brand. Michter's has been a big supporter of women. The brand tapped Pamela Heilmann, who had worked at Jim Beam, as its Master Distiller, a position she held until 2019. Currently at Michter's is Andrea Wilson, whose title is Master of Maturation. Wilson previously worked with the liquor giant Diageo, where she focused on whiskey planning, distillation, and maturation; helped revitalize the Stitzel-Weller facility; and managed production of Bulleit bourbon.

Nicole Austin got her start at Kings County Distillery in Brooklyn, later working for the Tullamore D.E.W. brand in Ireland, and found time in between to help start the New York State Distillers Guild and American Craft Spirits Association. In March 2018, she was hired as master distiller of George Dickel and has since reenergized that brand with a series of innovative bottlings. Heather Greene left the world of New York bourbon bars and the job of writing whiskey books to become an actual whiskey producer, making the much-praised Milam & Green brand bourbon and rye out of Texas.

Claire Marin, of Pollinator Spirits, makes bourbon in the Catskills. Becky Harris, co-owner and chief distiller at Catoctin Creek, produces a variety of ryes in Virginia. Lisa Roper Wicker is an itinerant whiskey distiller, perhaps best known for working at Widow Jane. Most recently, she joined Whiskey Thief Distilling in Frankfort, Kentucky, as Director of Distilling. Elizabeth McCall is the master distiller at Woodford Reserve. Marianne Eaves became Kentucky's first female bourbon master distiller since Prohibition, at Castle & Key. She is now an independent spirits consultant and the master distiller of Forbidden in Bardstown, Kentucky.

SOBA SOUR

NATASHA DAVID, RED HOOK, NEW YORK (2025)

Natasha David, a noted New York bartender who previously ran the Lower East Side bar Nitecap, and her husband, Jeremy, have a nightly ritual of sharing a pot of tea while they cook dinner. When Jeremy introduced a new soba (buckwheat) tea, Natasha knew she wanted to put it in a cocktail and the Soba Sour was born.

- 1½ ounces Soba Tea Bourbon (recipe follows)
- ¾ ounce fresh lemon juice
- ½ ounce Lustau Amontillado sherry
- ½ ounce Honey Syrup (see page 69)
- Dash of Fee Foam
- Orange slice and cocktail cherry, for garnish

Combine the liquid ingredients in a cocktail shaker half filled with ice and shake until chilled. Strain into an Old-Fashioned glass filled with one large ice cube. Garnish with the orange slice-cherry "flag" on a pick.

SOBA TEA BOURBON

- 750 ml bottle Elijah Craig bourbon
- 10 grams soba tea

Combine the soba tea and bourbon in a plastic container and let sit for 2 hours. Strain into a bottle.

DILEMMA

KAPRI ROBINSON,
ALLEGORY, WASHINGTON, D.C. (2024)

Kapri Robinson created this cocktail for a cocktail class she teaches. She told me, "I call it Dilemma because, though I am not a coffee drinker, I am very much into the flavors of coffee. I paired the coffee with one of my favorite ingredients: coconut."

2 ounces bourbon, preferably Bardstown

1/4 ounce Coffee Coconut Syrup (recipe follows)

2 dashes of chocolate bitters

2 dashes of Angostura bitters

Orange twist, for garnish

Combine ingredients in a mixing glass half filled with ice and stir until chilled. Strain into a rocks glass with one large ice cube. Garnish with an orange twist.

COFFEE COCONUT SYRUP

MAKES ABOUT 2 CUPS

80 grams (2 ounces) sweetened coconut flakes

80 grams (2 ounces) coffee beans

In a small saucepan over low heat, toast the coconut flakes until fragrant, about 2 minutes. Add 2 cups water and bring to a low boil. Add the coffee beans and let simmer for 10 minutes. Strain through a fine-mesh sieve, discarding the solids. Store in an airtight container in the refrigerator for up to 1 month.

CORE VALUES

KARIN STANLEY,
TATIANA, NEW YORK CITY (2025)

A cocktail maker of long standing in New York, Stanley told me this drink was an homage to NYC, using American rye and New York State apple distillate.

- 1½ ounces Rye & Sons rye
- 1 ounce Neversink Spirits Apple Apéritif
- ½ ounce Rhum J.M Terroir Volcanique
- ¼ ounce Brucato Amaro "Woodlands"
- Amarena cherry, for garnish

Combine the liquid ingredients in a rocks glass over one large ice cube and stir to chill. Garnish with the Amarena cherry.

2002

CRAFT DISTILLERIES

The story of bourbon and rye began with small distillers. Lots of them. The successful ones grew and became bigger distilleries. Then Prohibition came and only the strong survived. At the tail end, in 1934, only large distilleries remained. When the Kentucky Bourbon Trail was created in 1999, there were only nine distilleries making whiskey in the state. And that's how it stayed until the twenty-first century, when something called the craft distillery came along.

In 2002, New York State got its first new distillery since Prohibition. It was called Tuthilltown Spirits and was located in the Hudson Valley. The owners gutted and refitted a granary, bought a German pot still, took some courses (they didn't know how to distill whiskey), and got to work.

Until that year, they wouldn't have been able to embark on such a quixotic enterprise, because anyone who wanted to distill had to pay $50,000 for an industrial Class A license. But then the law changed and all you had to pony up was $1,450. The cocktail bars in New York City, with which the distilleries shared the term "craft," were happy to promote the new local booze. More craft distilleries quickly followed.

New York State's story played out similarly in other states. Colorado got Leopold Bros. Tennessee got Corsair. Pennsylvania got Dad's Hat. Texas got Balcones.

Many of these distilleries first put out a vodka or gin in order to bring in some money while they waited for their whiskey to age. Others aged their whiskey in smaller barrels than was the norm in order to speed along the aging process. And many released their whiskey while it was still quite young, often under two years, including Tuthilltown and Kings County Distilling, another New York distillery that garnered a lot of attention in the craft distilling movement's early years. Often these bottlings were 375 ml; yet the price was in line with a typical 750 ml. Small distilleries didn't have the big outfits' ability to buy and produce in bulk and thus keep prices down. Nonetheless, these small fry survived and thrived because of local boosterism, which proved to be a formidable force indeed.

Craft isn't a legal term. It has no real meaning; all distilleries employ craft in the making of their whiskeys. But it came to communicate certain notions, including small-scale production, a hands-on approach, a return to older practices, and the use of local ingredients. And these distilleries were indeed creative. Unlike hidebound liquor corporations, craft distillers experimented with unorthodox methods. Eventually that habit caught on. If your favorite big-name bourbon has come on with new expressions using unusual barrel finishes or mash bills, you can thank craft distillers. They probably did it first.

Today, there are around 2,700 craft distilleries operating in the United States. Of course, in the United States, where "grow or die" is in the business ether, you can't stay craft—that is, small—forever. Many of the early craft brands are now big business, having been bought up by bigger fish. In 2017, Tuthilltown was bought by William Grant & Sons, the Scotch distillery, which had already owned Tuthilltown's Hudson Whiskey brand since 2010.

RYE TAI

ALEX CLARK AND AMY GRINDELAND, FORT HAMILTON DISTILLERY, BROOKLYN (2023)

Fort Hamilton, founded in 2016, is a craft distillery located in the Industry City complex in Brooklyn. Like many modern microdistilleries, it runs an accompanying cocktail bar. This is one of the Fort Hamilton bar's signature drinks. The secret to this Rye Tai is to use both Double Barrel Rye and Cask Strength Rye to mimic the light and dark rum in the classic Mai Tai.

- 1½ ounces Fort Hamilton Double Barrel Rye
- ½ ounce Fort Hamilton Cask Strength Rye
- ½ ounce triple sec
- 1 ounce fresh lime juice
- ½ ounce orgeat
- Cocktail cherry, for garnish

Combine all liquid ingredients in a cocktail shaker half filled with ice and shake until chilled. Strain into an Old-Fashioned glass over one large ice cube. Garnish with the cherry and, if you have one, a tiny umbrella.

RED MOON

JOJO KITCHEN, SEATTLE, WA (2023)

Kitchen, a New Jersey native, is bar manager at Rob Roy, a prominent Seattle cocktail bar. This cocktail was created in 2023 for a local Seattle cocktail competition. A Sazerac riff using a local Washington State bourbon, it was originally inspired by the Lunar New Year, which was then the Year of the Rabbit.

- 1½ ounces Woodinville Straight Bourbon Whiskey
- ½ ounce Oloroso Sherry
- ½ ounce Coconut Pandan Syrup (recipe follows)
- Barspoon of apricot liqueur
- 10 dashes of Peychaud's bitters
- Absinthe, for rinsing
- Lemon twist, for garnish

Rinse the inside of a double Old-Fashioned glass with absinthe, discarding the excess. Set aside. Combine the whiskey, sherry, coconut syrup, apricot liqueur, and bitters in a mixing glass half filled with ice and stir until chilled. Strain into the prepared glass over one large ice cube. Garnish with the lemon twist.

COCONUT PANDAN SYRUP MAKES ABOUT 2 CUPS

- 520 ml coconut water
- 1 pandan leaf
- 780 grams sugar
- 3 grams salt or a pinch of salt

In a medium saucepan over low heat, warm the coconut water and pandan leaf. Do not let the mixture boil. Stir in the sugar until dissolved. Transfer to a container, let cool, and refrigerate. Leave the pandan leaf in the syrup overnight, then discard. The syrup will keep in the refrigerator for 1 month.

DELICIOUS APPLE HIGHBALL

MASAHIRO URUSHIDO,
KATANA KITTEN, NEW YORK CITY (2024)

Katana Kitten is a leading cocktail bar in Manhattan that does everything a little bit differently, all to pleasing ends. This highball is no exception.

1½ ounces Baby Jane bourbon	Build the drink in a highball glass filled with ice, adding the first three ingredients and then topping with soda. Garnish with the apple slice.
½ ounce Neversink Spirits Apple Apéritif	
½ ounce Martinelli's apple juice (reduced to ½ volume)	
Soda	
Apple slice, for garnish	

2002
OVERPROOF

In 2002, Buffalo Trace—upon the suggestion of a prescient customer—introduced George T. Stagg, an uncut and unfiltered powerhouse bourbon that was ahead of its time. It wasn't the first of its kind; Jim Beam had released Booker's in 1988. Still, this marked the advent of a coming mania for the intense experience afforded by overproof bourbons—that is, bourbons that are not cut with water but bottled at full alcoholic strength. Stagg is always bottled at no less than fifteen years' age and can exceed 70 percent alcohol; it is part of Buffalo Trace's "Antique Collection." It quickly became a collector's item (the original bottling went for a mere $40) and the limited run sells out every year. By 2012, at the San Francisco liquor store Cask, the waiting list for Stagg was 190 names long.

Soon, Stagg wasn't going stag on the shelves. High-alcohol spirits—variously sailing under the terms overproof, cask-strength, and barrel-strength—became commonplace. Scotches and bourbons dominate this bruising category.

In general, alcohol levels had been creeping northward for a few years, as distillers aimed to please bartenders and enthusiasts who thirsted for a more potent dram—50 percent had become the new 40 percent. But a good number of bottles now sped past even that mark, weighing in at anything from 51 to 70 percent, and beyond.

The appeal of these big guns is a mix of novelty and control. They represent something different, something that others don't have, as well as a way to try what in the past only the distillers got to taste when they went to the rickhouse and popped a few bungs. Once the bottle is home, moreover, buyers can play distiller and dilute—or not—to their heart's content.

Many advocates think overproof spirits simply taste better. Some of that flavor punch can arguably be attributed to the absence of chill filtering, a process that dispels fatty acids that cloud the liquid but can also remove flavor. It became common practice in the 1970s, when many distillers began lowering the standard alcohol level to 40 percent. But those acids aren't a problem in spirits above 50 percent, so high-alcohol distillates do not require chill-filtering.

Still, "cask-strength" doesn't necessary equal "better." It can also be a marketing tool, because it allows distillers to sell at a much higher price. Also, high-proof whiskey can also hide bad quality. But, as of this writing, the trend shows no sign of ebbing. If a distillery sells a bourbon or rye, chances are high that they also sell a cask-strength version of the same whiskey.

STAGGERAC

DON LEE,
PDT, NEW YORK CITY (2007)

This is a Sazerac riff with which to begin and end your evening. If you actually have some George T. Stagg on hand and want to drink like it's 2007—when the overproof bourbon was plentiful and relatively affordable—here's a cocktail for you. If you don't, you'll have to settle for drinking like it's 2022 and use Stagg Jr. Or you could try this recipe with a different overproof bourbon that is more gettable. But you'll have to change the name then.

2 ounces George T. Stagg or Stagg Jr. bourbon

Sugar cube

3 dashes of Peychaud's bitters

2 dashes of Angostura bitters

Absinthe, for rinse

Lemon twist, for garnish

Rinse a rocks glass with absinthe, discarding the excess. Muddle the sugar cube, bitters, and a barspoon of warm water at the bottom of a mixing glass. Add the bourbon. Add ice and stir until chilled. Strain into the prepared glass. Express the lemon twist over the drink and discard.

2002

FOUR ROSES

Four Roses was one of the great rediscoveries of the bourbon revival. For decades, the old brand was a no-show on American shelves. That's because Seagram bought the brand in the 1940s and, in the 1960s, decided to shift sales of the good straight bourbon to Europe and Asia. Meanwhile, it turned the American version of the flagship "yellow label" bourbon into a weak blend—because blended whiskey was all Seagram knew. In 1999, Seagram sold Four Roses, and it passed through several hands before landing with the Kirin Brewery, which decided to bring Four Roses back to the American market in 2002. (I have dated this entry with that year because, even though Four Roses is a brand that goes back to the 1800s, the Four Roses line we know today began when Kirin bought it.)

Jim Rutledge can be credited with much of Four Roses' revival. He began working at Seagram in 1966. Rutledge was a patient man. Once he found some ears that would listen, he convinced Kirin to bring Four Roses back to the United States and make it a real bourbon again. (Rutledge is about as plainspoken and direct a master distiller as you can imagine. He used to refer to the American Four Roses of the latter half of the twentieth century as a "bourbon-flavored alcoholic beverage.") As master distiller, a post he held from 1995 to 2015, he introduced well-regarded, single-barrel and small-batch expressions of Four Roses which, more than the benchmark "yellow label"—which remained a workhorse bourbon, just not a universally disrespected workhorse bourbon—did much to restore the sheen to Four Roses' name, particularly after they won a bunch of awards.

Four Roses is unusual among bourbon distillers in that it doesn't have one mash bill it works with, but two (one with a higher rye content, one with a lower), and it mixes these with five different yeast strains. That leaves the distillery with ten whiskeys to play around with and blend, making the Four Roses distilling process a bit of a Rubik's Cube—something for bourbon fans to play around with in their minds, imagining all the different flavors and combinations, whiskey-wise.

SELECTION
PS
Four Roses
SINGLE BARREL
Barrel Strength
MM
38-2M
SELECTION
PS
Four Roses
SINGLE BARREL
Barrel Strength
Kentucky Straight Bourbon Whiskey
Distilled and Aged by Four Roses Distillery LLC
Lawrenceburg, Kentucky
SS
22-1T

JIMMY RUTLEDGE

DON LEE, PDT,
NEW YORK CITY (2007)

This cocktail, from the early days of PDT, was named by Don Lee after the master distiller of Four Roses, Jim Rutledge. It is a spin on the classic cocktail, the Jimmie Roosevelt, which was a staple at Pegu Club, a bar Lee often haunted. Lee made it with the Four Roses Barrel Strength, which was just then released. "With the higher proof," Lee told me, "I felt it could stand up to being diluted with Champagne and the Chartreuse float." The drink was never on the PDT menu, but I was lucky enough to enjoy it as an off-menu item one evening and recalled it. When I asked Lee about it, he said, "You have one hell of a memory." It's one hell of a drink.

1½ ounces Four Roses Barrel Strength bourbon

2 ounces Moët White Star Champagne

¼ ounce green Chartreuse

Sugar cube

Dash of Angostura bitters

In a Champagne coupe, combine the sugar cube and the bitters. Add the bourbon and some cracked ice. Top with the Champagne. Float the Chartreuse on top.

2005

NEIGHBORHOOD COCKTAILS

Rye distillers across the United States owe a debt to Vincenzo Errico, an Italian who lived in New York City for just a few years in the aughts. He's made them a lot of money.

During his short time in the United States, Errico invented the Red Hook cocktail, in 2005. His aim was to create a modern riff on the old Brooklyn cocktail, which was composed of rye, sweet vermouth, Amer Picon, and maraschino liqueur. Errico, wanting to spotlight Italian products, kept the maraschino, but added Punt e Mes, a sort of cross between sweet vermouth and amaro. He began serving the drink at Milk & Honey, the trailblazing cocktail speakeasy where he worked. Customers liked it. But bartenders loved it. They'd never seen anything like it. A *new* rye cocktail? How was that possible? They'd been spending all their time re-creating *old* rye cocktails. Who knew you could take an old drink, sub out one or two ingredients, sub in something else of a similar sort, and end up with a new concoction that worked and tasted good? Science!

Soon there was a flood of new rye cocktails, all based on the Manhattan or the Brooklyn or both. They all followed the same basic format: rye whiskey, some kind of vermouth, and some kind of liqueur or bittering agent. There was the Greenpoint (rye, sweet vermouth, yellow Chartreuse), the Slope (rye, Punt e Mes, apricot liqueur), the Bensonhurst, the Cobble Hill, the Carroll Gardens, and on and on.

COBBLE HILL

SAM ROSS, MILK & HONEY, NEW YORK CITY (2005)

2 ounces Rittenhouse rye	In a mixing glass, muddle 3 of the cucumber slices. Add the liquid ingredients and half fill with ice. Stir until chilled. Strain into a chilled coupe. Garnish with the remaining cucumber slice.
½ ounce dry vermouth	
½ ounce Amaro Montenegro	
4 cucumber slices, 1 reserved for garnish	

CARROLL GARDENS

JOAQUÍN SIMÓ,
DEATH & CO., NEW YORK CITY (2008)

2 ounces Rittenhouse rye

½ ounce Punt e Mes

½ ounce Amaro Nardini

1 teaspoon Luxardo maraschino liqueur

Lemon twist, for garnish

Combine the liquid ingredients in a mixing glass half filled with ice and stir until chilled. Strain into a chilled coupe. Garnish with the lemon twist.

BUSHWICK

PHIL WARD,
MAYAHUEL, NEW YORK CITY (2009)

2 ounces rye

¾ ounce sweet vermouth

¼ ounce maraschino liqueur

¼ ounce Amaro Lucano

Combine the ingredients in a mixing glass half filled with ice and stir until chilled. Strain into a chilled coupe and serve without a garnish.

2005
RITTENHOUSE RYE

Prior to 2005 or so, few people knew of Rittenhouse rye except the people who made it. After 2005, every cocktail geek on the planet was a fan because all the cocktail bartenders and writers they looked up to were saying it was the best rye whiskey available (there weren't many) and the rye you should be using in your Manhattans and Boulevardiers.

Rittenhouse was a bonded rye, making it stronger and more flavorful than the other ryes on the market. It was also eight years old, much older than its competition. And so it ended up not only in classic cocktails but also in a lot of original cocktails invented in the aughts, including such noted potions as the Conference, Little Italy, and Carroll Gardens (see page 183). It was rare to see a cocktail menu during that decade that did not have at least one drink using Rittenhouse.

It was quite a reversal of fortune for the long-obscure brand. As can be deduced from the name, Rittenhouse was originally made in Pennsylvania. (It was originally named Rittenhouse Square.) It was made and launched by the Continental Distilling Company of Philadelphia in 1934. The rye was heavily advertised and was a big brand in Pennsylvania for many years. It began life as a young whiskey but became bonded in 1948.

Continental's founder, Harry Publicker, died in 1951 and the brand's stock tumbled after that. By the 1980s, Medley Distilling Company bought the brand and moved it to Kentucky. Rittenhouse changed hands a few more times before landing in 1993 with Heaven Hill, which relaunched the brand. Heaven Hill was making it when the rest of the world got wind of it.

The whiskey wasn't available in New York for a long time, but rye-loving cocktail historian David Wondrich did much to get the brand officially listed by the government liquor authorities of the state. After that, bars like Pegu Club made sure that every cocktail drinker knew about it. So did journalists. For the first seventy years of its existence, the only way Rittenhouse rye found its way into the newspaper was in an advertisement. Now, journalists dropped its name on a regular basis. Along with eight or nine other brands, it was one of the bottles upon which the cocktail and rye revivals were built.

BENSONHURST

CHAD SOLOMON,
MILK & HONEY, NEW YORK CITY (2006)

This drink is yet another entry in the Neighborhood Cocktail sweepstakes, and one of the best.

2 ounces Rittenhouse rye

1 ounce dry vermouth

2 teaspoons maraschino liqueur

1 teaspoon Cynar

Combine the liquid ingredients in a mixing glass half filled with ice. Stir until chilled. Strain into a chilled coupe.

2007
MAD MEN

It was the good fortune of the AMC series *Mad Men,* which debuted in 2007, to unveil its drink-centric world at a time when a growing fraternity of alcohol enthusiasts was rediscovering America's rich drinking history. Cocktails were a vital element of the show right from the opening scene, which showed protagonist Don Draper sitting in a bar. Before the audience learns his name or his profession, he expresses his drink preference: "Do this again—Old-Fashioned, please."

That many ad men drank deeply seems unquestioned. When I was working on an article about *Mad Men* for *The New York Times,* the bartender and bar lore archivist Brian Rea, who worked in the 1950s at the Little Club, a popular Midtown restaurant, told me, "Lunch was a big thing. They took two and a half hours. We had a lot of agency people come in, from Cunningham & Walsh, BBDO, all having serious lunches with drinks." Carlo Marioni, a New York bartender with more than forty years' experience who last worked at Pietro's, agreed: "Those years, for lunch, they used to drink three Martinis. Then they'd come back before dinner for Rusty Nails, White Spiders."

Draper's preference for Old-Fashioneds did its share in bringing not only that classic cocktail back into fashion but also the bourbon and rye that serve as the base of the drink. In a critical scene in season three, we see Draper—who is dodging a work-related Kentucky Derby party by hiding in the country club's bar—build an Old-Fashioned from start to finish, using a bottle of Old Overholt. "You look like you know what you're doing," observes his unexpected guest at the bar (who turns out to be Conrad Hilton). He did. And because he did, many of us now do, too.

MID-CENTURY MUDDLED OLD-FASHIONED

This is a standard formula for a muddled-style Old-Fashioned. If you really like your orange and cherry, you can muddle one of each at the bottom of the glass *and* garnish the drink with the same.

2 ounces bourbon

2 dashes of Angostura bitters

Soda water

Sugar cube

Orange slice, for garnish

Cocktail cherry, for garnish

Muddle the sugar cube, bitters, and a bit of soda water in the bottom of an Old-Fashioned glass. Add the bourbon and stir briefly. Add ice and stir again. Garnish with the cherry and the orange slice.

2008

DAVE PICKERELL

You may have noticed quotes from a man named Dave Pickerell in this book (below and on page 134). That's because Pickerell, who died in 2018, was once everywhere. The Ohioan earned his stripes as a distiller at Maker's Mark, where he worked for fourteen years, ending in 2008. After that, he went rogue, becoming an itinerant talent-for-hire and peddling his vast distilling knowledge to whoever would listen and pay for it. He saw the changing landscape of American whiskey before many others did. There were new distilleries that needed expertise and new ideas. Pickerell was the encouraging uncle to all these whiskey wannabes.

He's been called the Johnny Appleseed of the modern whiskey world, and the comparison is apt. Among the new American whiskeys where he was on the ground floor were Whistlepig, Hillrock, J. Rieger & Co., Corsair, and Garrison Brothers. He was the guy who discovered the golden cache of Canadian rye whiskey that made Whistlepig such a success right out of the gate.

His work on Hillrock was emblematic of his innovative ways. Hillrock's inaugural offering was a bourbon. No big deal there. But it was a solera bourbon. In the solera system of aging, a succession of barrels, as they are tapped, are topped off with younger spirits once or twice a year. Liquids from different years intermingle and marry, creating a whiskey that blends different flavors into a kind of average, which makes for a more consistent product on shelves. It's a practice that's usually associated with the sherry industry, where it has been standard procedure for years. Pickerell didn't see why, with some practical problem-solving, it couldn't be applied to bourbon. In selecting the aging casks, he turned poetic. "Generally, soleras are sherries finished in bourbon barrels," he told me. "We decided to have a little fun with it and finish ours in sherry barrels." Voilà: a sherry solera turned on its head.

Hillrock further distinguished itself by building its own malting house. Malting is the practice by which dampened, germinating barley is dried and thus halted in its growth; the grain can then proceed to fermentation. A handful of Scotch producers still malt a portion of their barley, but the practice is all but extinct in the United States.

Pickerell looked the part of your friendly traveling whiskey distiller. He wore jeans and had a big belly and a small cowboy hat. But he was a modern thinker. He also didn't shun the cocktail world, as many whiskey distillers did, and seemed to enjoy going to booze conventions and cocktail bars and jawing with the press. You'd run into him at bars the way you never would with other distillers, unless you had an appointment.

Pickerell's last major project was creating Blackened, a brand of blended straight whiskey, for the heavy metal band Metallica. He aged it in brandy barrels, with Metallica music played in the background to provide the liquid with the appropriate vibrations. So, the whiskey rock star finally worked with actual rock stars.

2010s

WHISKEY WRITERS

Whiskey Writer. Who ever heard of such a person?

In 2000, no one had. As a profession, the line didn't exist. And then, suddenly, it did.

By 2015, the shelves at the local bookstore were groaning with whiskey books. Soaring consumer interest in American whiskeys over the previous decade had created a symbiotic cottage industry in writing about those spirits. Unlike almost any other spirits—Scotch is a notable exception—bourbon and rye and other American whiskeys had been taken up by many writers as their primary, and sometimes only, topic.

Fred Minnick, a former Army photographer who wears pastel jackets and ascots, emerged as the king of the genre. "I liken covering American whiskey to covering sports," said Minnick in a 2016 *New York Times* article I wrote profiling the profession. "Sports teams have these very rabid fans. Bourbon fans are the same way. They want to know every single thing." Other practitioners who worked this seam more or less full time included Lew Bryson (*Tasting Whiskey: An Insider's Guide to the Unique Pleasures of the World's Finest Spirits*), Clay Risen (*American Whiskey, Bourbon & Rye: A Guide to the Nation's Favorite Spirit*), Robin Robinson (*The Complete Whiskey Course*), Noah Rothbaum (*The Whiskey Bible*), Heather Greene (*Whiskey Distilled: A Populist Guide to the Water of Life*), and Aaron Goldfarb (*Hacking Whiskey*).

None of these journalists worried about running out of copy.

"American whiskey is complex," Minnick said in the same article. "Every day there is someone acquiring a little guy, or there is a lawsuit. From that side of things, there's a lot of drama. If you have drama, you have something to write about."

The very circumscribed nature of American whiskey lends itself to exacting reportage, Bryson said. "The problem with writing about rum is there's no regulation, and it changes and does whatever it wants," Bryson said. "But whiskey is quite regulated, and there is a lot more expectation there from the drinkers. You can make predictions and statements about it."

With so many on the trail of the same stories, you may expect some backbiting. But most journalists describe their interactions with the same word: "congenial." Within reason, of course.

"I see it as akin to the bourbon distillers down in Kentucky," said Rothbaum. "They're all friendly, which is not to say people aren't competitive."

UPWARD SLOPE

AARON GOLDFARB,
BROOKLYN (2025)

"Despite being a long-time bourbon writer and drinker, people are often surprised to learn I'm not a huge fan of Manhattans and rarely drink them," said Goldfarb, who created this cocktail for this book. "That's not the case, however, with Julie Reiner's The Slope, by far the best Manhattan variant in my opinion. The Slope calls for apricot liqueur as a modifier, which can, unfortunately, be hard to find. That's what led me one day to try and swap in something much cheaper and more readily available—DeKuyper Peachtree Schnapps. Believe it or not, in 1987, it was one of the bestselling alcohols in the entire country, utilized in cocktails like the Sex on the Beach and the Fuzzy Navel. It was so popular, in fact, the Jim Beam company thought they should diversify their portfolio by offering DeKuyper's parent company, National Distillers, $545 million dollars to acquire the brand; National Distillers was also more than happy to throw in several bourbons they likewise owned that absolutely no one was drinking during this clear-spirits era—Old Grand-Dad, Old Crow, Old Taylor, and Old Overholt Rye."

- 2 ounces Old Grand-Dad Bonded Bourbon or another bonded bourbon
- ½ ounce Old Overholt Bonded Rye or another bonded rye
- ½ ounce Punt e Mes
- ¼ ounce dry vermouth
- ¼ DeKuyper Peachtree schnapps
- 2 dashes of Angostura bitters
- Lemon twist, for garnish

Combine the liquid ingredients in a mixing glass half filled with ice and stir until chilled. Strain into a chilled coupe. Garnish with an expressed lemon twist.

2010s

BOULEVARDIER

Throughout history, bourbon cocktails have sported many flavors. One thing they were not, however, was bitter. That changed in the 2010s, when bartenders rediscovered the Boulevardier, a lost 1920s cocktail that is simply a Negroni made with bourbon. Once found, the drink spread widely, giving bars another vintage bourbon vehicle that went beyond the Old-Fashioned and the Manhattan.

The drink's pedigree is American, albeit in exile. It was created in the late 1920s by Erskine Gwynne, a reckless and high-living expatriate writer, socialite, and Vanderbilt relation. Gwynne was born in Paris and later founded a Paris literary magazine called the *Boulevardier*. The monthly covered "the lighter side of life in Paris" and ran from 1927 to 1932. Mr. Gwynne was evidently a regular at Harry's New York Bar, a Paris watering hole favored by Americans. He advertised the *Boulevardier* magazine in the pages of *Barflies and Cocktails,* a book published by the bar's owner Harry MacElhone. In the ad he described his publication as the sort to be read "before, between and after cocktails." Perhaps as a thank you, MacElhone included the Boulevardier cocktail as a footnote in the back. Ironically, it was a bourbon cocktail invented at a time when legal bourbon wasn't being made, except for medicinal purposes.

There it sat for roughly eighty years. Why the sudden popularity during the cocktail renaissance? It helps that the drink is only a short sidestep from the trendy Negroni, which asks for gin instead of whiskey. Also, like the Negroni, it doesn't require a degree in mixology to make or understand. Finally, it contains Campari, and mixologists like any cocktail with a bitter note.

In recent years, bartenders have tended to substitute rye for the bourbon.

BOULEVARDIER

TOBY CECCHINI,
THE LONG ISLAND BAR, BROOKLYN (2014)

Beginning in the late aughts, Cecchini became a big advocate for the revival of the Boulevardier cocktail. When he opened Long Island Bar in 2014, he made it the bar's signature cocktail. Unlike the original, which called for bourbon, Cecchini uses rye.

- 1 ounce Old Forester Rye
- 1 ounce Old Overholt Bonded Rye
- 1 ounce Campari
- ½ ounce Carpano Classico vermouth
- ½ ounce Cinzano sweet vermouth
- Orange twist, for garnish

Combine the liquid ingredients in a mixing glass half filled with ice and stir until chilled. Strain into a chilled coupe. Garnish with the orange twist.

2010
WHITE DOG

In the late twentieth century, the bourbon industry put out clear versions of its spirits, denuded of color, because it was desperate to compete with the prevailing vodka and tequila crazes.

In the early twenty-first century, bourbon once again started putting out clear liquor, but this time not out of desperation. It was seizing an opportunity, capitalizing on the public's growing thirst for anything bourbon-related. They marketed it as white dog. White dog, or white whiskey, is, basically, moonshine. It's newborn whiskey, crystal-clear grain distillate as yet unkissed by the barrel.

By 2010, white dog was having its day. In that year, Kris Comstock, a representative for the Buffalo Trace distillery in Kentucky, conducted a seminar on bourbon at Char No. 4, a bar in Cobble Hill, Brooklyn, that offered 150 kinds of American whiskey. "The first thing that everyone wanted to taste was the white dog," he said, in an article I wrote about white dog for *The New York Times*. "We make products that win amazing awards all around the world, and they want to taste the white dog."

Aging whiskey in wood has many beautiful effects, but it does tend to disguise whatever the base spirit is. When you strip that away, you're getting a real sense of what corn offers, or rye or wheat, in terms of flavor. Unlike vodka, in which the source grain is often purposefully purified to a vanishing point, white dogs are pungently fragrant, with a chewy sweetness to them.

Most of the first white dogs came from young microdistilleries like Death's Door in Wisconsin; Finger Lakes Distilling in upstate New York; Tuthilltown in the Hudson Valley; the Copper Fox Distillery in northern Virginia; and House Spirits in Portland, Oregon. They were all looking to bank some cash while waiting for their whiskeys to mature. But the big boys soon got in on the game. Buffalo Trace began a commercial release of its white dog. Soon, there were so many white dogs on the market that some bars offered a white-whiskey flight. The trend was fueled by two very differently motivated groups: practical-minded distillers and ravenously questing enthusiasts.

Some observers wondered if the distilleries would continue selling white dog when all their other whiskeys had matured. As it turned out, no. Only a few commercial white dogs still survive today. But they didn't disappear before a new lesson in the making of American whiskey had been taught. White dog was no longer a taste only known to distillers.

SPICY STRAWBERRY MARGARITA

HERMAN C. MIHALICH,
FOUNDER AND DISTILLER, DAD'S HAT
PENNSYLVANIA RYE WHISKEY (2025)

Dad's Hat, located just outside Philadelphia, was one of the forerunners in the rebirth of Pennsylvania-style rye whiskey. It is also one of the only such distillers to produce a white dog rye.

- 2 ounces Serrano-Infused White Rye (recipe follows)
- 1 ounce strawberry puree
- ¾ ounce fresh lime juice
- ½ ounce agave syrup
- Pinch of salt
- Lime wheel, for garnish

Combine all ingredients except the garnish in a cocktail shaker half filled with ice. Shake until chilled. Strain into a rocks glass filled with ice. Garnish with the lime wheel.

SERRANO-INFUSED WHITE RYE

- 750 ml bottle Dad's Hat White Rye
- 3 serrano peppers

Score the peppers and let steep in the white rye for 24 hours. Strain the rye, discarding the peppers, and store in a clean bottle at room temperature. It will last indefinitely.

2011
INDIANA

The state everyone talks about when they discuss bourbon is Kentucky. The states everyone used to name-check when they talked about rye were Pennsylvania and Maryland. And yet, when you look at the back label of many bottles of bourbon and rye, you will see that they are distilled in Indiana. Specifically, Lawrenceburg, Indiana.

Lawrenceburg is a town of five thousand people on the banks of the Ohio River, just miles from the borders of both Ohio and Kentucky. The town is not littered with distilleries pumping out dozens of brands of bourbon and rye. No, it has only one. But that one is a giant. It has the most unsexy name in all American whiskey: Midwest Grain Products. But everyone calls it MGP, which is somehow even more boring.

MGP makes whiskey for people who don't have the ability to make whiskey themselves—because they don't have a distillery yet, because they don't have enough stock in their rickhouses, or for other reasons. MGP used to be a dirty secret in whiskey circles. No distiller wanted anyone to know they were slipping across state lines and asking their Hoosier neighbor if they could borrow a cup of whiskey. But that's what increasingly happened in the modern age of the non-distilling producer (see page 142), when lots of newbies wanted to cash in on the bourbon and rye craze but were years away from being able to make their own juice. Among the now-famous brands that initially sourced their liquid from MGP are Bulleit, High West, Templeton, New Riff, Pinhook, James E. Pepper, and Smooth Ambler. A lot of big bourbon and Tennessee Whiskey outfits also turned to MGP when they decided they wanted a rye whiskey, including George Dickel. MGP is like Home Depot; it's the first place people shop when they're going to build something new.

The location always made sense. It was next to the Ohio River, allowing for easy transport of its whiskey, and Lawrenceburg sits on that same big underground limestone shelf that Kentucky benefits from in terms of water filtration.

MGP may be a whiskey distillery at its most industrial, but it has come to be respected for some of its product, particularly the 95 percent rye mash bill it produces, which has been bottled under many names. MGP also puts out a few of its own brands, including George Remus and Rossville Union. Curiously, these are among the least popular of the whiskeys MGP makes.

RHAPSODY IN RYE

TOM MACY AND JULIE REINER, BAR 54, NEW YORK CITY (2014)

Macy and Reiner came up with this rye Manhattan variation for a midtown Manhattan rooftop hotel bar where they were consultants for a time.

- 2 ounces rye
- ¾ ounce dry vermouth
- ½ ounce Ramazotti
- ½ ounce Bénédictine
- Dash of Angostura bitters
- Dash of orange bitters
- Orange twist, for garnish

Combine the liquid ingredients in a mixing glass half filled with ice. Stir until chilled. Strain into a chilled coupe. Garnish with the orange twist.

2011

FANCY WHISKEY BARS

The bourbon and rye boom fostered several adjacent cottage industries. One was the whiskey bar—that is, bars whose focus was whiskey, almost to the exclusion of all other spirits. Multnomah Whiskey Library, which opened in 2013 in Portland, Oregon, was a typical specimen. It has a lot of the things you'd expect to find in a whiskey bar—leather chairs, dark wood, a certain amber glow. But it also has something you don't see every day: a ladder.

The bartenders needed those steps to reach some of the higher-perched spirits in its collection of 850 whiskeys. In fact, this new breed of whiskey parlors could be called "ladder bars" because they possess such an enormous variety, far beyond that of the average whiskey bar, that every square inch of wall space must be put to work.

You found ladders, too, at Hard Water, a waterfront bar in San Francisco that specialized in rare and vintage whiskeys; Canon, in Seattle, where roughly two-thirds of the 3,500-bottle collection are whiskeys; the Flatiron Room in Manhattan, which carries 750 to 1,000 whiskeys at any given time; and Jack Rose Dining Saloon, the Washington, D.C., bar that can be considered the father of the über whiskey bar, having opened in 2011.

These saloons arrived at a time when there was suddenly much more whiskey to stock. Distilleries in Kentucky, Scotland, Ireland, Japan, and other whiskey strongholds were releasing many more varieties than they had in the past: small batch, single barrel, barrel proof, limited edition, etc. That's a lot of whiskey. And obsessive bar owners are intent on cramming as much as they can into one space, because their equally obsessive customers will buy them.

Surprisingly, given their huge stock, most of these bar owners say few bottles gather dust. "The vintage stuff only sells about every two months or so, but everything else moves pretty well," Jamie Boudreau, the owner of Canon, told me for a 2014 *New York Times* article I wrote on the subject. "The one beautiful thing about the collection is that because people order so many different things, most bottles usually last four to six months before needing to be replaced."

These ungainly collections didn't grow overnight. The owner of Multnomah hired a "curator" a year ahead of time to assemble the bar's collection. Mr. Boudreau had been collecting bottles for a decade before opening Canon.

The bars' vast collections tend to perpetuate themselves. "Every penny Canon has made goes right back in the business," Mr. Boudreau said. When Canon opened, the Captain's List, his name for the full spirits menu, was five to six pages. Now, it's nearly two hundred.

TRIUMPH

TOMMY TARDIE, THE FLATIRON ROOM, NEW YORK CITY (2014)

A former bartender at The Flatiron Room—one of the best whiskey bars in Manhattan—made this drink for owner Tardie, knowing he generally preferred his alcohol neat. He loved it and it became his favorite. The bartender named it Triumph for the motorcycle Tardie drove.

1½ ounces Wilderness Trail bourbon	Combine the liquid ingredients in a mixing glass half filled with ice. Stir until chilled. Strain into a rocks glass filled with one large ice cube. Garnish with the orange twist.
½ ounce Pedro Ximénez sherry	
2 dashes of chocolate bitters	
Orange twist, for garnish	

2013
TENNESSEE WHISKEY

If you want to start an argument in an American whiskey bar, bring up the subject of Tennessee Whiskey and take a strong stand on what it exactly is: bourbon by another name, or its own thing.

The laws about the category are confusing. On a federal level, Tennessee Whiskey need only be "a straight bourbon whiskey authorized to be produced only in the State of Tennessee." That's not enough for the state of Tennessee, though. In a piece of 2013 legislation, it insists that all Tennessee Whiskeys undergo a filtering process that is called, rather poetically, the Lincoln County Process. This is where the whiskey is filtered through (or steeped in) a thick layer of maple charcoal before it goes into barrels. This has a smoothing effect on the whiskey that you will immediately notice if you compare a Tennessee Whiskey to a bourbon. To my mind, the charcoal filtering is enough for Tennessee Whiskey to merit its own category. It is a distinction *with* a difference. (There's a 1981 country hit called "Tennessee Whiskey," first recorded by David Allan Coe, then George Jones, then Chris Stapleton. Its refrain is, "You're as smooth as Tennessee whiskey.")

Truthfully, there wouldn't be much of a debate about this if one of the biggest whiskeys in the world, Jack Daniel's, didn't insist on calling itself a Tennessee Whiskey. The other major player in Tennessee Whiskey, the Pepsi to Jack Daniel's Coke, is George Dickel, which has its own faithful following and has lately, under master distiller Nicole Austin, shown itself to be the more innovative of the two.

Dickel's history is complicated. George Dickel was an actual man, a German immigrant who settled in Tennessee in the mid-nineteenth century, but he was a wholesaler, not a distiller, selling whiskey made in Cascade Hollow. The Dickel track record as a consistent product has been spotty over the decades, with all kinds of interruptions. (Tennessee enacted Prohibition a decade before the nation did and kept it for a decade longer.) The most recent was around the turn of the century, when international giant Diageo acquired the brand and plant and seemed not to care. It was in mothballs for years, not reopening until 2003. This, ironically, made people care about Dickel again, mainly because they thought they wouldn't be able to get it anymore.

In recent years, Tennessee Whiskey has benefited from the increased interest in bourbon and rye. Whereas there were once only two examples of the style, there are now several, including Prichard's, Nelson's Green Briar, and Uncle Nearest (see page 211). Not to be undone by its northern neighbor, the Tennessee Whiskey Trail was founded in 2017.

DINKY TROPICALE

KITTY BERNARDO,
THE DINKY BAR & KITCHEN, PRINCETON, NEW JERSEY (2016)

"I've wanted to use Jack Daniel's in a tropical cocktail for a while because Jack has a very distinct banana note on the finish that always stood out to me when I was a young bartender. I thought making a tropical drink with Jack was so weird and out there as a concept at the time," Bernardo told me. "The fernet is there because of the minty finish it adds to the drink, cutting through the rich coconut cream."

1 ounce Jack Daniel's Tennessee Whiskey

2 ounces Coco Lopez Cream of Coconut

1½ ounces pineapple juice

½ ounce Fernet Branca

¾ ounce Goslings Black Seal Rum

Orange wheel, for garnish

Freshly grated nutmeg, for garnish

Combine all ingredients except the Goslings and garnishes in a cocktail shaker filled with ice. Shake until chilled. Strain into a goblet or snifter filled with cracked ice. Float Goslings on top. Garnish with the orange wheel hugging the rim of the glass and with freshly grated nutmeg.

2013

CELEBRITY WHISKEYS

The bandwagon's journey went like this. Bourbon and rye were getting hot again in the aughts. So, some young, brash entrepreneurs started founding craft distilleries, hoping to cash in on the craze. Some were self-made booze enthusiasts. Others were tech and finance cast-offs looking for another, more interesting line of work after the financial crash of 2008. Then came the even more wealthy, often backed by hedge-fund forces, who opened distilleries that were big and tricked-out for tourism and brand-spanking new and shiny from the get-go. These outfits often sold out to even bigger boys after five to ten years of doing business.

Lastly came the celebrities. George Clooney showed all the actors and artists in the world that you didn't have to slave away making art to get rich. He founded Casamigos Tequila in 2013 and sold it four years later for $1 billion. Most of the bold-face names that followed his example also placed their bets on tequila and sometimes mezcal. But a few attached their names to whiskey. There were athletes (Terry Bradshaw, Charles Woodson, Scottie Pippen, Stephen Curry), actors (Matthew McConaughey, Jamie Foxx), and musicians (Drake, Brad Paisley). But the biggest—and most unlikely—star by far was Bob Dylan, whose Heaven's Door bourbon and rye came out in 2018.

2014

UNCLE NEAREST

Fawn Weaver, an entrepreneur, investor, and former PR executive, was having breakfast on the top floor of the Four Seasons hotel in Singapore when she read a June 25, 2016, story in *The New York Times* by Clay Risen. Like much of the rest of the world, that article was Weaver's first introduction to Nathan "Nearest" Green—Jack Daniel's first master distiller and the man who taught Daniel how to distill.

"A lot of journalists do this," Weaver told me. "They don't have the time or resources or budget with their various publication [to follow up on a story], so they do a 'lob.' They find out as much as they can and write it in such a way, in hopes that someone will complete the work. Usually no one picks it up." Weaver fielded Risen's lob and ran with it. She moved to Tennessee, did exhaustive research on Green; convinced Brown-Forman, which owns Jack Daniel's, to formally recognize Green's contribution; bought the farm where Green and Daniel first worked together; built the distillery; and, by 2017, produced the whiskey that she named after Green. This made Weaver the first African American woman to head a major spirits brand, and she did it with an all-female executive team.

And just like that, America's most famous whiskey, and the bourbon industry in general, was—just like many fields of enterprise in American history—forever connected to the country's still-unresolved history with slavery. Weaver basically rewrote the history of America's most famous whiskey brand. Since then, other famous bourbon brands have acknowledged, with varying success and honesty, their past reliance on the labors and talents of enslaved or formerly enslaved Black Americans.

"When Uncle Nearest debuted, it was the first spirit brand to commemorate an African American," Weaver told me in an interview for my newsletter, *The Mix*. "Since then, dozens have launched, helping to bring a greater awareness of inclusivity to our industry. The bourbon industry was built by both whites and African Americans, but the latter were lost to history. Returning them to the foundation of the industry opens it up to the 70 percent of the country that consider themselves minorities, women, and people of color, and who embrace brands that celebrate all of America versus just 30 percent."

"Nearly every bourbon company in Kentucky had African American distillers," she continued, "yet we don't know the names of any of them. My hope is the large bourbon companies will dig through their files, pull out those names, and shine a light on them." (In 2025, the Uncle Nearest distillery, facing financial difficulties, went into receivership.)

GROWN FOLK MIMOSA

FAWN WEAVER,
UNCLE NEAREST DISTILLERY,
SHELBYVILLE, TENNESSEE (2021)

This twist on the brunch classic comes from the founder of Uncle Nearest Tennessee Whiskey herself.

½ ounce Uncle Nearest 1884 Small Batch Whiskey	Combine all ingredients except the Champagne in a cocktail shaker. Shake until chilled. Strain into a flute. Top with the Champagne.
½ ounce Cointreau	
2 ounces orange juice	
1½ ounces Champagne	

2022

THREE-CHAMBER STILL RYE

In 2021, when Leopold Bros., an experimental distiller in Colorado, started sending out samples of its Three Chamber Rye Whiskey, nobody had ever tasted a rye whiskey made with a three-chamber still. That's because they were alive. The only people who knew how such whiskey was made or was supposed to taste were mostly dead, because people stopped using three-chamber stills to make rye several decades ago.

Todd Leopold had long been interested in the old techniques of making spirits. His bookworm ways eventually led him to a paper written by C. A. Crampton and L. M. Tolman in 1907 for the Bureau of the Commissioner of the Internal Revenue. Leopold gleaned from this document that most of the rye producers polled by Crampton and Tolman were distilling their juice using something called a three-chamber still, a device not used by anyone in the current American whiskey industry. Once he had a grasp of what the still was, he commissioned still makers Vendome to construct one. Vendome agreed on the condition that there was no guarantee that it would work. The resulting still was twenty feet tall and four feet in diameter, with four separate chambers.

Then Leopold went in search of a rye strain that would have the same low starch content as those used pre-Prohibition. Modern commercial ryes, which are bred for high yield, don't have as much flavor as they did a century ago. He settled on Abruzzi rye and began growing it.

"You don't have to understand how the still works to understand that this is an entirely different kind of whiskey where the grain matters," said Leopold. "The barrel isn't dominating the finished whiskey."

The Three Chamber Rye was indeed a challenging brew when Leopold released it in 2021. Did it taste like old American rye? Who really knew? But it certainly tasted different, grassy and floral and bright. It was also hard to find—there wasn't much of it. Accordingly, it was also expensive. Fortunately, a development was around the corner that would make it more palatable and accessible.

"He called me," Nicole Austin, the master distiller at George Dickel, told me for an article I wrote about the subject for *The Mix*. "His ultimate goal with creating the three-chamber rye was really to restore this historical spirit, and his research shows that it was commonly blended with a column-distilled rye. To truly achieve that final goal of restoring this spirit style he needed to obtain a column-distilled rye. He called me asking me to find him some. And I said, 'Actually, I have some.'"

Austin created the base Dickel rye blend and shared it with Leopold. He did the same. They started doing some blending of the two bases. Blending samples were mailed back and forth between Tennessee and Colorado. Austin won't say what the blended ratio between the two whiskeys is, because it's based on a sensory standard, not a number. "If I do more of this, I want to match the standard, so the ratio may change."

The whiskey began as a limited release. Soon after, it became a regular line extension. With that, one could arguably consider the return of classic East Coast rye whiskey to the American scene to be complete.

TULLAHOMA SOUR

DAVID WONDRICH, CREATED FOR GEORGE DICKEL (2021)

Booze historian David Wondrich created this cocktail for the Dickel-Leopold Bros. spirit collaboration, naming it after Dickel's hometown.

- 1¾ ounces George Dickel x Leopold Bros Collaboration Blend
- ¾ ounce fresh lemon juice
- 1 rounded teaspoon granulated sugar
- ¼ ounce kümmel
- ¼ ounce orange curaçao
- Brandied cherry, for garnish (optional)

Briefly stir the lemon juice and sugar in a cocktail shaker. Add the whiskey, kümmel, and curaçao. Shake well with ice and strain into a chilled cocktail glass. Garnish with the brandied cherry, if desired.

2026

PRESENT DAY

And that brings us to the present day. What will the future of bourbon and rye bring? Current headlines focus on the bourbon bubble. After a quarter century of growth, it has burst, leaving distillers faced with waning interest and a potential glut of aging stock. As of fall 2025, American whiskey production was down to its lowest level since 2019. In late 2025, Jim Beam announced plans to pause production at its main distillery for a year. What will this mean for the bourbon and rye drinker? Will prices fall? Will treasured brands and distilleries disappear? Will the next generation benefit from hyper-aged whiskey looking for an audience? Will whiskey writers need to find a new beat?

Who can tell? The stories of bourbon and rye in America—like the tumultuous story of the nation itself—have never lacked for drama. There's always another chapter and it's usually an unexpected one. Whatever it may be, rest assured there will be a good bottle of bourbon or rye nearby to toast it with.

ACKNOWLEDGMENTS

I'd like to express my heartfelt thanks to all the paid subscribers of my Substack newsletter, *The Mix with Robert Simonson,* and everyone who attended the Martini Expo in Industry City, Brooklyn, September 12–13, 2025. As much as anyone, your steady support made the completion of this book possible.

Thanks to all the bartenders, bar owners, and writers who contributed recipes to this book, including (in alphabetical order), Dave Arnold, Nicole Austin, Brian Bartels, Kitty Bernardo, the late Dick Bradsell and Brother Cleve, Bea Bradsell, Jackson Cannon, Erick Castro, Toby Cecchini, Chuck Cerankosky, Gary Crunkleton, Charlotte Daniel, Natasha David, Dale DeGroff, Meaghan Dorman, John Dye, Steve Ferreira, St. John Frizell, Aaron Goldfarb, Amy Grindeland and Alex Clark, Charles Hardwick, Chris Hannah, Kurt Hernon, Lynn House, Charles Joly, Allen Katz, JoJo Kitchen, Jeff Knott, Don Lee, Tom Macy, Lindsay Matteson, Herman C. Mihalich, Jeffrey Morgenthaler, Aaron Post, Julie Reiner, Kapri Robinson, Leo Robitschek, Sam Ross, Jon Santer, Audrey Saunders, Francis Schott, Joaquín Simó, Todd Smith, Chad Solomon, Karin Stanley, the late Murray Stenson, Tommy Tardie, Sother Teague, Masahiro Urushido, Mike Vacheresse, Thad Vogler, Phil Ward, Fawn Weaver, and David Wondrich.

I am grateful to the talented American whiskey writers who were generous enough to make the time to sit down and talk with me about bourbon and rye: Aaron Goldfarb, Heather Greene, Clay Risen, Robin Robinson, and Noah Rothbaum. I recommend the books written by these people.

Also, thanks to the generous Joe Magliocco, owner of Michter's, who kindly offered his thoughts and time.

Thanks to my editor, Julie Bennett; the book's designer, Annie Marino; and the staff at Ten Speed Press, who were patient with me when this book took longer than expected to complete. Special thanks to photographer Lizzie Munro. This is the fourth cocktail book on which we have collaborated.

Eternal thanks to my agent, David Black, for his continual support and faith.

Love to my son, Asher, and stepson, Richard. And, lastly, love and thanks to my wonderful wife, Mary Kate Murray, who is not only my life partner but was also a partner in this book, helping to collect all the fantastic cocktail recipes you find herein.

INDEX

Page numbers in *italics* refer to images.

TEN SPEED PRESS
An imprint of the Crown Publishing Group
A division of Penguin Random House LLC
1745 Broadway
New York, NY 10019
tenspeed.com
penguinrandomhouse.com

Typefaces: Jeff Levine Font Foundry's Classroom, Sharp Type's Tarnac and Post Grotesk, , Hoefler & Co.s Sentinel, ITC's Franklin Gothic

Library of Congress Cataloging-in-Publication Data is on file with the publisher.

Hardcover ISBN 978-1-9848-6333-1
Ebook ISBN 978-1-9848-6334-8

Editor: Julie Bennett | Production editor: Terry Deal
Designer and art director: Annie Marino
Production designers: Mari Gill and Faith Hague
Production and prepress color manager: Jane Chinn
Copy editor: Jonathan Milder | Proofreaders: Rachel Whitten, Miriam Taveras | Indexer: Eldes Tran
Publicist: David Hawk | Marketer: Chloe Aryeh

Manufactured in China

10 9 8 7 6 5 4 3 2 1

First Edition

The authorized representative in the EU for product safety and compliance is Penguin Random House Ireland, Morrison Chambers, 32 Nassau Street, Dublin D02 YH68, Ireland, https://eu-contact.penguin.ie.